Artificial Intelligence

Originally published in 1992, this title reviews seven major subareas in artificial intelligence at that time: knowledge acquisition; logic programming and representation; machine learning; natural language; vision; the design of an AI programming environment; and medicine, a major application area of AI. This volume was an attempt primarily to inform fellow AI workers of recent European work in AI. It was hoped that researchers in 'sister' disciplines, such as computer science and linguistics would gain a deeper understanding of the assumptions, techniques and tools of contemporary AI.

T0179142

Artificial Intelligence

Research Directions in Cognitive Science:
European Perspectives Vol.5

Edited by
D. Sleeman and N.O. Bernsen

Routledge
Taylor & Francis Group

LONDON AND NEW YORK

First published in 1992
by Lawrence Erlbaum Associates Ltd

This edition first published in 2020 by Routledge
2 Park Square, Milton Park, Abingdon, Oxon OX14 4RN
605 Third Avenue, New York, NY 10017

First issued in paperback 2021

Routledge is an imprint of the Taylor & Francis Group, an informa business

Publisher's Note
The publisher has gone to great lengths to ensure the quality of this reprint but points out that some imperfections in the original copies may be apparent.

Disclaimer
The publisher has made every effort to trace copyright holders and welcomes correspondence from those they have been unable to contact.

ISBN 13: 978-0-367-40605-9 (pbk)
ISBN 13: 978-0-367-40596-0 (hbk)

Artificial Intelligence

Research Directions in Cognitive Science: European Perspectives Vol 5

Edited by

D. Sleeman

*Computing Science Department,
King's College, The University,
Aberdeen AB9 2UB*

and

N.O. Bernsen

*University of Roskilde, Institute 7, P.O.
Box 260, 4000 Roskilde, Denmark*

LEA LAWRENCE ERLBAUM ASSOCIATES, PUBLISHERS LEA
Hove (UK)   Hillsdale (USA)

Lawrence Erlbaum Associates Ltd., Publishers
27 Palmeira Mansions
Church Road
Hove
East Sussex, BN3 2FA
U.K.

British Library Cataloguing in Publication Data

Artificial intelligence. - (Research directions in
   cognitive science. ISSN 0961-7493; 5)
   I. Sleeman, D. II. Bernsen, Niels Ole III. Series
   153

   ISBN 0-86377-176-9

The text in this book was produced direct from disks supplied by the authors, via a
desktop publishing system. Printed and bound by BPCC Wheatons Ltd., Exeter, UK

Contents

Acknowledgements

The several anonymous referees who provided valuable feedback on earlier drafts of the papers.

Dr Peter Edwards, Ruediger Oehlmann and Fengru Chen (Aberdeen) provided invaluable help in preparing the final manuscript.

The Computing Science Department at Aberdeen which provided some technical, secretarial, and financial support for this venture.

Artificial Intelligence: Achievements and Promises*

D. Sleeman
Computing Science Department, King's College, The University, Aberdeen AB9 2UB, Scotland UK

This introduction reviews the toplevel goals of the field, and outlines some of the techniques evolved in the three decades of AI's existence. Some of the field's remaining challenges are discussed. This introduction draws its principal illustrations from the European work which appears in the ensuing chapters.

1. INTRODUCTION

In this introduction, I attempt to give an overview of the field's goals, the techniques and methodologies which have evolved during the three decades of the field's existence, and some achievements to-date.

There have been many attempts to define the field and to specify its research methodologies. An early definition which I still believe captures much of the field's spirit is due to Minsky:

> AI is the science of making machines do things that require intelligence if done by men. (Minsky, 1962)

Humans have a wide range of capabilities, but ranking high in importance is our ability to solve problems and to learn and retain information acquired during problem solving. The bad news for AI and for Cognitive Psychology is that there is neither a general theory of problem solving nor a general theory of human learning[1]. Theories of human learning generally hold only for a narrow range; problem solvers

---

*This introduction is based on Sleeman (1989).

[1] The good news is that AI provides psychology with a strong methodology and a powerful set of tools with which to develop and refine such theories.

are thus forced to use a series of ad hoc heuristics when solving any non-trivial task. These factors make AI an empirical science.

During the three decades of its existence AI has pursued in parallel, several (complementary) research goals; these can be summarised as:

Building intelligent machines (or artifacts);
Encoding in machine-manipulatable form (the whole of) human expertise;
Understanding human cognition.

These objectives make it clear that AI is essentially an interdisciplinary subject with particularly close relationships to Psychology, Philosophy, Linguistics and Logic[2]. However, it must be stressed that Computer Science has a central position because of the need to implement and test a *model* of the phenomena being investigated. Model building is central in AI as it is in a number of other empirical sciences (such as Operations Research). Below we list two sets of ill-defined/ill-specified tasks:

Predicting changes in a country's economy
Calculating throughput for a factory
Weather forecasting
The physiological function of the liver

A grammar for, say, English
Speech understanding
How students learn specific topics
Medical diagnosis
Engineering design
Scientific discovery

The first set of tasks have been approached by formulating mathematical models; the second set have been approached by AI which to-date has been (extensively) concerned with *symbolic* models. However, the process of model building and verification are essentially the same whether the models are mathematical or symbolic.

1.1   Techniques Evolved by AI

We saw earlier the very lofty goals which the field had set itself. In this section we will review some of the field's solid, and indisputable, technical achievements. Below, we summarise the major achievement of the field to date:

---

[2]AI specialists also often need to understand something of the subject domain in which they are working, e.g., medicine, biochemistry, etc.

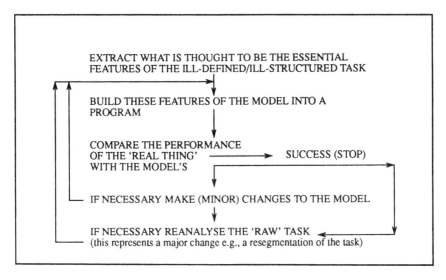

EXTRACT WHAT IS THOUGHT TO BE THE ESSENTIAL
FEATURES OF THE ILL-DEFINED/ILL-STRUCTURED TASK

BUILD THESE FEATURES OF THE MODEL INTO A
PROGRAM

COMPARE THE PERFORMANCE
OF THE 'REAL THING'   →   SUCCESS (STOP)
WITH THE MODEL'S

IF NECESSARY MAKE (MINOR) CHANGES TO THE MODEL

IF NECESSARY REANALYSE THE 'RAW' TASK
(this represents a major change e.g., a resegmentation of the task)

FIG. 1.   Building computer models for complex tasks.

Representation
    Predicate Calculus
    Production Rules
    Semantic Networks
    Frames
    Data-structures of LISP and PROLOG
Search: Systematic and Heuristic.
Inference Techniques: Deduction
Learning
Natural Language Understanding
Perception and Robotics
Programming Languages and Environments.

Each of these topics will be considered in a subsection.

1.1.1   Representation

The task to be considered by the computer, by the program, *has* to be
represented in some form. Thus, before information from a visual scene
can be searched for features, it has to be represented in a program; for
example, an image from a TV camera may be stored in the elements of
a 1000x1000 array, where each element represents the intensity of the
corresponding part of the image. Although representation is of *crucial*
importance to AI, only a limited number of representational schemas
have been defined. These include:

Predicate Calculus (Robinson, 1965), Production Rules (Newell, 1973), Semantic Networks (Quillian, 1968), Frames (Minsky, 1975) and, one could argue, the data structures available in AI's most commonly used languages, i.e. LISP and PROLOG. See Barbuti, Martelli and Simi (this volume) and Halpern (1986) for more detailed discussions of these issues.

Figure 2 shows how frames can be used to represent knowledge about a particular house, garden and family. Each frame has a series of named "slots". In the case of the "house" frame these are: address, storeys, rooms, construction, owner, and *pointers* to family and garden frames. The idea of a default value is an important one; this is the slot's value unless it is explicitly changed.

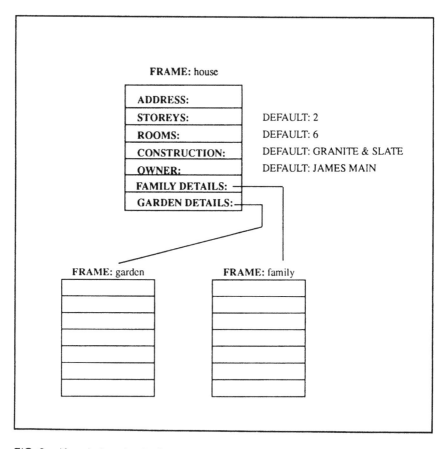

FIG. 2.   Knowledge about a house represented as a frame.

A further important issue with representation is the *granularity,* i.e. the level of detail which is kept. For instance, it is clear a housewife, an architect, a rate assessor and a plumber will all wish to know something about the plumbing of a house; but will all *need* information at a variety of different levels of detail.

1.1.2 Search

Strange as it may seem search, *systematic* search, is one of AI's major techniques. Newell (1969) refers to *search* as being a weak method, that is the *default* method which can be used (by humans and programs) when a more specific, more powerful algorithm is not available. Figure 3 shows the *initial* state, *goal* state and *operators* (moves) for solving the 8-puzzle. Figure 4 shows the search tree which can be created for this particular problem (Nilsson, 1971).

In case you should think that only *trivial* tasks can be solved by search, I list some non-trivial tasks which have been solved with this technique:

Transforming algebraic and trigonometric expressions from one form to another;

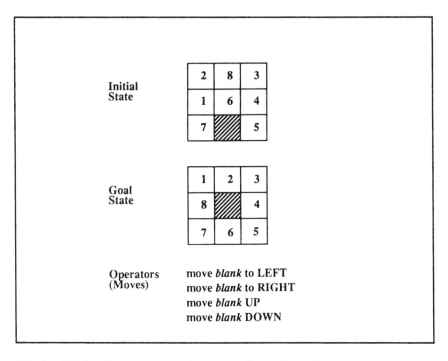

FIG. 3.   Initial state, goal state and operators for the 8-puzzle.

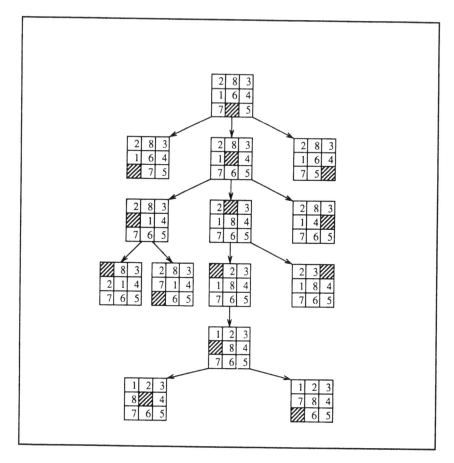

FIG. 4.   A solution tree for the 8-puzzle using the operators defined in Fig. 3.

Generating a possible synthesis path for organic chemistry;
Parsing natural language sentences;
Determining moves in Draughts/Checkers;
Determining moves in Chess.

Figure 5 gives details for search spaces for Trigonometry and Chemical Synthesis. To stress the complexity of the tasks undertaken it should be noted that Sridharan (1973) pointed out that his program SYNCHEM was able to solve tasks which had taken teams of the world's best synthetic chemists several years to solve.

Domains which have a sizeable number of operators, will produce sizeable search spaces. Hence, the need to determine the more *promising* parts of the search-space. This has led to the development of techniques for *pruning* search trees and *heuristics* (rules of thumb) to decide which are the more promising parts to be explored. Sometimes these heuristics are general, and other times they are very domain specific; for example, all nodes which mention "blood" should be discarded. (Our current state of knowledge about search techniques is partly grounded in theory, but is also partly reliant on empirical results (Nilsson, 1971).)

1.1.3 Inference Techniques

AI has developed most fully deduction. Humans can deduce consequences from a series of facts. If told that John has a daughter Mary, and Mary's son is Peter, then we can deduce that John is Peter's grandfather. Similarly, if told that John is taller than Mary and Mary is taller than Peter, humans can deduce that John is also taller than Peter.

Considerable effort was expended in the 1960s in developing computational inference algorithms for the Predicate Calculus (Robinson, 1965). Subsequently, these insights (algorithms) have been

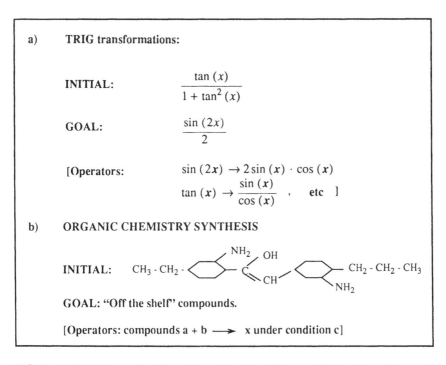

FIG. 5.   Initial state, goal state and operators for a)trignometric transformations and b) the organic chemistry synthesis task.

embodied in the PROLOG programming language (Colmerauer, et al. 1973). Humans also use a wide range of other techniques including analogy, negative inference (making inferences from what is *not* known), and island building (working both *forwards* from an original task and *backwards* from an *assumed* solution). All of these techniques have been incorporated in AI systems. (Inference and Analogy are discussed briefly in the next paragraph.)

1.1.4   Learning

Learning has been seen from the beginning of AI as a *critical* capability; if machines could not learn then AI's credibility would be questioned *and* moreover, its value as a technology would be greatly limited. To date, there have been two major approaches to learning within symbolic AI, namely what is now called Similarity Based Learning (SBL) and Explanation Based learning (EBL). If a SBL system is told that {dog, cat, cow} are members of the class mammal, such algorithms would attempt to infer the set of characteristics which the examples have in common; often several descriptors are suggested. EBL systems, on the other hand, require a complete domain theory, and are able, on the basis, of a *single* example to explain which features are relevant to its class membership. So the fact that a dog is warm-blooded is significant, but the fact that it is called "Joey", lives in central Paris, and wears a Diamond-studded collar are not. (The process underlying analogy are clearly related to rule inference).

Kodratoff (this volume) gives a detailed review of recent European work in learning.

1.1.5   Natural Language Understanding

Much of human communication is through Natural Language, and hence the importance of machines having a deep competence in this area. In the early 1950s Turing suggested that we might consider a machine to be intelligent if a human operator could not decide whether he was communicating through a terminal with a human or with a computer, (Turing 1963).

Historically, Natural Language work has been a major impetus to the development of further representational schemas. More recently, AI has realised that *inference* is integral to the correct understanding of many sentences. Extensive real world knowledge is often required to understand sentences which are syntactically similar. For example, this is needed for a proper understanding of "they" in the following sentences:

The government stopped the student demonstration because *they* were afraid of violence.

The government stopped the student demonstration because *they* advocated revolution.

Bunt (this volume) reviews some of the recent (European) theoretical work in Linguistics.

1.1.6 Perception and Robotics
Robots which can both "see", "hear" and analyse speech will find many practical uses in our factories and homes. Further, it is clear that before this technology emerges, it will be necessary for the field to solve a number of challenging scientific as well as technical problems. See Nagel (this volume) for a review of recent work in Vision.

1.1.7 Programming Languages and Environments
In the introduction (section 1), we highlighted the centrality to AI's research methodology of model building and model evaluation. As models, programs, are often discarded or modified after only a few runs, it is important that these programs can be developed and debugged rapidly. Hence the need within AI for high level programming languages and programming environments which provide good diagnostic and editing facilities when programs malfunction. Smith, Sloman and Gibson, (this volume) discuss this issue. Additionally, AI needs programming languages which allow programs to be self modifying; this was a novel feature when first introduced in LISP by McCarthy.

1.2   A Historical Review of AI

Having reviewed some of the basic themes and techniques of AI, I will now try to give some perspective on how the field has evolved over the last two decades or so.

I have chosen to divide the overview given in Figure 6 into 5 phases: Ad hoc, Axiomatic/formal, Procedural approaches, Knowledge-Based Systems and Multiple Approaches. In the ad hoc phase, workers were attempting to demonstrate, by any means, that a program could emulate intelligent behaviour so it was particularly important to demonstrate that programs could learn and communicate in natural languages[3].

The Axiomatic/formal phase arose as a reaction to the ad hoc data-structures and inference/deductive procedures used in the first phase. In this phase, workers tried to show the advantages of using a standard representation and inference/deduction mechanism based on predicate calculus. Indeed, during this phase a considerable number of

---

[3]Both capabilities are demanded by the Turing test.

I: ADHOC (1957-65)

GENERAL PROBLEM SOLVER; (Newell & Simon, 1963)
GAME PLAYING: CHECKERS; (Samuel, 1963)
BASEBALL: QUESTION-ANSWERING SYSTEM; (Green et al., 1963)
GEOMETRIC-ANALOGY PROGRAM; (Evans, 1968)

II: AXIOMATIC/FORMAL (1964-68)

Algorithmic procedures for Predicate Calculus; (Robinson; 1965)
STRIPS: ROBOT PLANNING; (Fikes & Nilsson, 1971)
INTEGRATED ROBOT; SRI, 1970s; (Fikes, Hart & Nilsson, 1972)
SEARCH; (Nilsson, 1971)

III: PROCEDURAL APPROACH (1968-74)

PLANNER; (Hewitt, 1971)
MICRO-PLANNER; (Winograd, 1972)
SHRDLU (NATURAL LANGUAGE PROGRAM); (Winograd, 1973)

IV: KNOWLEDGE-BASE SYSTEMS (1970-)

DENDRAL; (Buchanan & Feigenbaum, 1978)
MYCIN; (Shortliffe, 1976)
R1; (McDermott, 1982)

Intelligent Tutoring Systems:

DEBUGGY; (Brown &Burton, 1978)
LMS/PIXIE; (Sleeman, 1982)

V: Multiple approaches (1978-)

LEARNING; (Mitchell, 1978; Michalski, 1980)
REPRESENTATION AND MANIPULATION OF MULTIPLE INTERACTING
KNOWLEDGE SOURCES; (Erman et al., 1980).
CONSTRAINT SATISFACTION; (de Kleer, 1986).

FIG. 6.   A historical review of Artificial Intelligence.

technical advances were made, to enable the formalisms of 19th century mathematics to be used as a computational tool. However, a number of problems were soon noted with this approach. Namely, it was extremely hard to represent some tasks in Predicate Calculus and secondly that some, even simple tasks, lead to combinatorial explosions when run with the standard inference "engine". Proponents of the third phase, the so called Procedural Approach, argued that a more powerful and indeed appropriate representation for AI systems was procedural knowledge.

In this formalism even a simple fact like:

John has red hair

was represented as a procedure.

At about this time (early 1970s), the Stanford group noted that one of the things which distinguished experts from novices in a field was *knowledge.* They noted that the emphasis in early work on AI had been on (clever) algorithms, say, for doing search, and that these algorithms were often only provided with minimal task-specific knowledge, for example, the famous Missionaries and Cannibals problem. At the heart of the Stanford concern, was whether algorithms which could cope with these "toy" domains could cope with the knowledge used by a doctor to diagnose a rare heart condition, or that used by an engineer when designing a motor car engine. So in summary, the concerns addressed during this phase were: how to acquire domain expertise, how to encode this knowledge in an intelligent system as well as which algorithm(s) would be appropriate to interpret sizeable Knowledge Bases.

As the field has matured, it has realised that the several techniques and approaches formulated above, (i.e. predicate calculus, procedural and knowledge-based) are appropriate for different *types* of tasks, and so the fifth phase reflects this. Additionally, under this heading I have included some of the major research themes which are currently being explored.

Both work in Natural Languages and Intelligent Tutoring Systems have made it clear that much of the data processed by AI systems is both incomplete and inconsistent. AI initially assumed that its systems would only process complete and consistent information. This was a reasonable initial assumption which has since been shown to be false. (Science often makes strong simplifying assumptions when tackling new problems).

1.3 AI's Current Research Agenda

Many of the themes noted earlier (section 1.1) under AI's achievements, reappear here, too. This is natural. Some basic building blocks (e.g. Representational Schemas) were defined at an early stage, but are proving to be inadequate now that a range of more complex projects are being undertaken, Lenat and Guha (1990).

Some of the achievements noted were subfields (like Natural Language and Robotics) in which progress has been solid, but where major challenges remain. However, more specialized forms of problem solving have evolved in the last decade, namely, techniques for propagating domain constraints, as well as Qualitative ("Common-sense") Reasoning. Additionally, given significant "real-world" tasks have been attempted by AI, several new techniques are now included in its standard toolbox, including dealing with uncertainty; techniques for integrating information from a number of knowledge sources; and integrating database technology with AI techniques (e.g., accessing large rule-sets within a knowledge engineering tool).

European work in AI has contributed and continues to contribute to these important themes. Additionally, we have noted the importance of engineering good environments which can help with the rapid building and debugging of Computer-Based models.

1.4   Exploiting Today's AI technology: Expert Systems

Expert Systems illustrate nicely the well-known maxim:

"Today's (research) problems are tomorrow's products".

Expert Systems represent just part of today's AI technology — the exploitable part of the technology. Below we summarise the principal characteristics of Expert Systems:

Expert systems are AI programs as they use symbolic information and heuristic inference procedures;
High performance (expert-like) in narrow areas of expertise;
Flexible: easy to add knowledge to the system;
Can cope with noisy data;
Understandability — can explain their solutions (as can an expert).

Stefanelli (this volume) reviews a number of expert systems which have been built in the area of Medicine using today's technology. Additionally, Reichgelt and Shadbolt (this volume) describe a knowledge based system for Knowledge Acquisition which is poised to make an impact on the ubiquitous Knowledge Acquisition bottleneck. These two chapters demonstrate the depth and breadth of European activities with this new technology.

1.5   The Field's Research and Development Strategy

I stressed earlier that in the absence of strong theories of problem solving, AI is forced to be an experimental, empirical, science. Hence, the field and European programs in this area, pursue a range of Development, R&D as well as Basic Research activities:

Level 1: Use well understood technologies/techniques on tasks which appear superficially to be similar to those for which they had previously succeeded. (For example, we used an Expert System shell to diagnose diseases in Scots pines where the technology had previously been used to diagnose human diseases.)

Level 2: Use tried technologies/techniques on a range of analogous but more demanding tasks; "stretching" the technology. For example, create a rule base of a 1000 rules, where it had only previously been run with toy example(s), to check that the technique can be "scaled" up.

Level 3: Projects which we have no clear idea how to solve, namely "Basic" Research, for example Reasoning by Analogy and Qualitative Reasoning.

I would like to make two additional points. Firstly, the solution of level 1 and level 2 tasks do frequently throw up Basic Research issues. Secondly, Basic Research issues can themselves be undertaken in the context of important real-world tasks. For example, Qualitative Reasoning Research is being undertaken in the context of Medicine and Petroleum Geology as well as in the context of what are disparagingly known as "toy problems". The AI Researcher needs to specify clearly the formal problem s/he wishes to pursue, and then to search, and often search hard, for a real-world domain in which this can be meaningfully explored.

The above paragraphs argue the case for a multifaceted approach to research methodology in a pragmatic field such as AI which lacks a strong theoretical underpinning. Further, we noted earlier the need for interdisciplinary approaches to AI problems, citing the relevance of Computer Science, Cognitive Science, and Linguistics, to building AI systems. Workers in each of these disciplines need to understand the fundamental nature of the assumptions, methods and techniques of the other fields. This volume attempts primarily to inform fellow AI workers of recent European work in AI; moreover, given the general theme of the series of volumes, it is hoped that Researchers from our "sister" disciplines will also find it deepens their understanding of the assumptions, techniques and tools of contemporary AI.

REFERENCES

Barbuti, R., Martelli, M. & Simi, M. (This volume) Knowledge Representation and Logic Programming.

Brown, J.S. & Burton, R.R. (1978). Diagnostic models for procedural bugs in basic mathematical skills. *Cognitive Science, 2,* pp 155-192.

Buchanan, B.G. & Feigenbaum, E.A. (1978). DENDRAL and meta-DENDRAL: Their applications dimension. *Artificial Intelligence, 11.*

Bunt, H. (This volume) Language Understanding by Computer: Developments on the Theoretical side.

Colmerauer, A. et al (1973). *Etude et Realisation d'un systeme PROLOG Convention de Research IRIA-Sesori No 77030.*

Erman, L.D., Hayes-Roth, F., Lesser, V. & Reddy, D. (1980). The HEARSAY-II speech-understanding system: Integrating knowledge to resolve uncertainty. *Computing Surveys, 12, 2,* pp 213-253.

Evans, T.G. (1968). A Program for the solution of Geometric-Analogy Intelligence Test. In *Semantic Information Processing* (ed M.L. Minsky). Cambridge, MA: MIT press.

Fikes, R.E. & Nilsson, N.J. (1971). STRIPS: A new approach to the Application of Theorem Proving to Problem Solving. *Artificial Intelligence, 2,* pp 189-208.

Fikes, R.E., Hart, P.E. & Nilsson, N.J. (1972). Learning and Executing Generalised Robot Plans. Artificial Intelligence, 3, pp 251-288.

Green, B.F., Wolf, A.K., Chomsky, C. & Laughery, K. (1963). BASEBALL: An automatic question answerer. In *Computers and Thought,* (ed. E.A. Feigenbaum & J. Feldman). New York: McGraw-Hill.

Halpern, J. (1986). Proceedings of Conference on *Theoretical Aspects of Reasoning about Knowledge.* Los Altos: Morgan-Kaufmann.

Hewitt, C. (1971). PLANNER: A Language for Proving Theorems in Robots. *Proceedings of IJCAI-71 conference.*

de Kleer, J. (1986). An Assumption-based TMS. *Artificial Intelligence, 28,* pp 127-162.

Kodratoff, Y. (This volume) Characterising Machine Learning Programs: A European Compilation.

Lenat, D.B. & Guha, R.V. (1990) *Building Large Knowledge-based Systems.* Menlo Park, California: Addison-Wesley.

McDermott, J. (1982). R1: A Rule-based Configurer of Computer Systems. *Artificial Intelligence, 19,* pp 39-88.

Michalski, R.S. (1980). Knowledge Acquisition through Conceptual Clustering: A Theoretical Framework and an Algorithm for Partitioning Data into Conjunctive Concepts. *Policy Analysis and Information Systems, Vol 4, 3,* pp 219-244.

Minsky, M.L. (1962). Problems of formulation in the artificial intelligence area. *Proceedings of a Symposium on Mathematical Problems in Biology.* New York: American Mathematical Society.

Minsky, M. (1975). A Framework for Representing Knowledge. In P. Winston (Ed.). *The Psychology of Computer Vision.* New York, NY: McGraw-Hill.

Mitchell, T. (1978). Version Spaces: An approach to concept learning. PhD Dissertation, Department of Computer Science, Stanford University. (Technical Report STAN-CS-78-711).

Nagel, H-H. (This volume) A Perspective on Machine Vision.

Newell, A. & Simon, H.A. (1963). GPS: A Program That Simulates Human Thought. In E.A. Feigenbaum & J. Feldman (Eds.) *Computers and Thought.* New York, NY: McGraw-Hill.

Newell, A. (1969). Heuristic Programming: III-Structured Problems. In J. Aronofsky (Ed.) *Progress in Operations Research, III.* New York, NY: Wiley.

Newell, A. (1973). Production systems: Models of control structures. In W. Chase (Ed.), *Visual Information Processing.* New York, NY: Academic Press.

Nilsson, N.J. (1971). *Problem-Solving methods in Artificial Intelligence.* New York: McGraw-Hill.

Quillian, R. (1968). Semantic Memory. In M Minsky (Ed.) *Semantic Information Processing.* Cambridge, MA: MIT Press.

Reichgelt, H. & Shadbolt, N. (This volume) ProtoKEW: A knowledge based system for Knowledge Acquisition

Robinson, J.A. (1965). A Machine-Orientated Logic Based on the Resolution Principle. *J. ACM, 12,* pp 23-41.

Samuel, A.L. (1963). Some studies in Machine Learning using the game of Checkers. In *Computers and Thought* (ed. E.A. Feigenbaum & J. Feldman). New York: McGraw-Hill.

Shortliffe, E.H. (1976). *Computer-based medical consultations: MYCIN.* New York: Elsevier.

Sleeman, D.H. (1982). Assessing competence in basic algebra. In *Intelligent Tutoring Systems* (edited by D Sleeman and J S Brown). London: Academic press, pp 185-199.

Sleeman, D.H. (1989), Artificial Intelligence: Fact or Fiction? *Aberdeen University Review, 182,* pp 136-157.

Smith, R., Sloman, A. & Gibson, J. (This volume) POPLOG's two-level virtual machine support for Interactive Languages.

Sridharan, N.S. (1973). "Search strategies for the task of Organic Chemical Synthesis". *Proceedings of IJCAI-73 conference,* pp 95-104.

Stefanelli, M. (This volume) Applications of Expert Systems technology to Medicine.

Turing, A. (1963). Computing Machinery and Thought. In *Computers and Thought,* (Feigenbaum, E.A. & Feldman, J. eds). New York: McGraw-Hill.

Winograd, T. (1972). *Understanding Natural Language.* New York: NY: Academic Press.

Winograd, T. (1973). A Procedural Model of Language Understanding. In R.C. Shank and K.M. Colby (Eds.) *Computer Models of Thought and Language.* San Francisco, California: Freeman.

CHAPTER 1

Knowledge Representation and Logic Programming

Roberto Barbuti
Dipartimento di Informatica, Università di Pisa, Italia

Maurizio Martelli
Dipartimento di Matematica, Università di Genova, Italia

Maria Simi
Dipartimento di Matematica e Informatica, Università di Udine, Italia

1. INTRODUCTION

The underlying assumption of most of the current research in Artificial Intelligence is that intelligent systems can be constructed using explicit, declaratively represented, factual knowledge together with general reasoning mechanisms. Therefore the study of formal ways of extracting information from symbolically represented knowledge is recognized of central importance in the field of Knowledge Representation (KR); see (Brachman & Levesque, 1985) for a large collection of papers in the field.

In the history of Knowledge Representation, a great deal of effort has gone into the study and development of various notations and formalisms, with rather less attention paid to logical foundations and inferential consequences of the notations. However many authors have argued about the fundamental importance of a formal semantics to give a precise account of the meaning of the notations used (Hayes, 1974; McDermott, 1978) and defended the role of formal logic as an essential tool for modelling and understanding representation formalisms (Hayes, 1977; Moore, 1982). This point of view has now become widely accepted.

When First Order Predicate Logic is used as a representation formalism, the basic problem of deciding whether a fact is a logical consequence of a knowledge base is in general unsolvable and intractable. Most techniques and formal proposals in Artificial Intelligence can be understood as attempts to deal with only limited kinds of reasoning in exchange for reasonable complexity of behaviour. The explicit recognition of this "fundamental trade-off in knowledge representation and reasoning" is, for example, the basis of current research on terminological languages and hybrid systems (Levesque & Brachman, 1984).

The need of a powerful theorem prover for full First Order Logic has influenced many early studies in improving the efficiency of the proof procedures: from the seminal work on resolution of Alan Robinson (Robinson, 1965) to the important results on SLD-resolution (Kowalski & Kuehner, 1971; Hill, 1974). The realization that a subclass of First Order Logic (Horn clauses) has a procedural interpretation (Kowalski, 1974), together with some experiments on theorem provers for natural languages (Colmerauer, 1973), gave rise to the PROLOG language and to a new field called Logic Programming (LP); (see (Lloyd, 1987; Apt, 1987) for an introduction to the theory; (Sterling & Shapiro, 1986) for a nice presentation of the language and its applications. Many important foundational aspects of Logic Programming can be found in (Kowalski, 1979a, 1979b; van Emden & Kowalski, 1976; Apt & van Emden, 1982; Clark & Tarnlund, 1982)).

Logic Programming is characterized by two essential features: non-determinism, i.e. search-based computation, and unification. The central role of search in LP languages is connected with their being halfway between theorem provers and standard programming languages, which makes them suitable for many artificial intelligence applications.

Programs are statements in a logical language and a program execution is a proof from the set of axioms (the program) of a specific formula (the goal). As observed in (Pereira, 1985), peculiar to LP, with respect to automated deduction, are deduction steps which are simple so that their effect will be easier to predict (as for functional application). The boundaries between LP and automated deduction is not a sharp one and there is always a mutual influence; new results in automated deduction can widen what is considered LP and results in LP can improve the field of automated deduction.

First Order Predicate Logic has other problems as a tool for KR; in fact, classical logic is inadequate to deal with common sense reasoning problems, where one has to draw conclusions based on incomplete knowledge or defaults and has to reason about knowledge and belief and other mental attitudes, time, actions and plans. Current research in these fields remains within the logic framework, originating a number of non-standard logics that try to capture the essential aspects of our every-day way of reasoning about the world. Recently in the LP community several extensions have been proposed in order to cope with such representation problems.

Knowledge representation and logic programming are getting together (as it was foreseen by Kowalski in (Kowalski, 1979a)), especially in recent years when, from one side the knowledge representation community has evolved towards formal representations with concerns about efficiency, and on the other side the logic programming community has developed many powerful extensions and integrations with other programming paradigms, which make logic programming languages increasingly able to cope with the representation of complex knowledge.

A possible direction of development for LP is a system which enables the user to write specifications using the full power of logic with the ability to refine them into efficient logic programs (for example see (Bundy, 1988)). This kind of approach will make the difference between LP and KR even smaller.

In this paper we survey some research topics and results in KR and we emphasize the recent proposals in LP which can be considered as contributions to these research topics. The aim is to show that research in LP, with its necessary extensions and improvements, can contribute a formal and efficient representation of knowledge that emerges from the recent research in KR.

In Section 2 we will discuss some classical AI paradigms and proposals to accommodate them into a logical framework. These proposals have been taken into consideration as possible LP applications since the early years of LP. Section 3 is devoted to the new trend of KR towards hybrid systems and specialized reasoners and of LP towards new extensions and integrations. From the KR point of view the emphasis is on using different and specialized reasoning components, while from the LP point of view the aim is to offer a richer language which integrates different paradigms. These approaches can be seen as part of the same trend towards multiple paradigm systems resulting in similar solutions in the two fields. Section 4 covers an important area of KR research, i.e. non-monotonic reasoning, which is producing a number of different non-standard logics to deal with common sense reasoning tasks, and is also playing a more and more important role in LP, especially for the treatment of negation. Section 5 addresses the problem of reasoning about knowledge and belief pointing out how one of the approaches coming from KR, i.e. meta-level reasoning, is also becoming usefully applied in LP. Finally, in Section 6, we outline some research directions towards higher-order logic which will play a significant role in providing a unifying framework for research proposals in KR and LP.

2. STRUCTURED REPRESENTATION OF KNOWLEDGE

This section will briefly review some of the more popular paradigms for representing knowledge which emerged from the AI community, namely *production systems* or rule based representations, associative representations such as *semantic networks* and object centred or *frame based* representations. Each of these paradigms can be understood as a class of proposals able to cope with a number of specific representation and reasoning tasks. We will discuss early attempts to include these proposals in the LP framework.

2.1   Production Systems, Semantic Networks, Frames

Production Systems

Production rules are one of the most popular paradigms for representing knowledge in expert systems. This is due to the fact that in some tasks it appears that the domain knowledge of an expert can be conveniently expressed as a set of "if-then" associations, called rules, of the form *"if* antecedent *then* consequent".

A production system can be considered as constituted of three parts: a rule base, a working memory and a rule interpreter. The interpretation cycle proceeds essentially as follows: the antecedents of all rules are matched against every element in the working memory and the set of rules that are applicable is determined (the conflict set). A rule in the conflict set is then activated according to a conflict resolution strategy. The activation changes the working memory by executing the consequent part of the rule and the cycle repeats.

A disadvantage of this representation schema is that it forces the user to think about control of the inference process by selecting the most appropriate conflict resolution strategy. This can be embedded in the interpreter or encoded in meta-rules.

The expressive power is in general very weak but nevertheless adequate for a significant number of applications. Techniques for speeding up the computation of the conflict set have been developed by exploiting a special internal representation of the rules (Forgy, 1982) and parallel processing (Gupta, 1985). This makes possible the development of applications involving thousands of rules.

Semantic Networks

Semantic networks, originated by Quillian (Quillian, 1967), are a model accounting for the organization of "semantic" knowledge (as opposed to syntactic) in the human mind. Quillian emphasizes the connectivity of knowledge; like in a dictionary, each concept is defined in terms of others, creating a linked structure connecting all the concepts of interest. As a data structure, a semantic network is constituted of nodes, corresponding to concepts (individual entities or classes of entities), and directed arcs corresponding to binary relationships between concepts.

Among these relations, the sub-class relation between concepts (often called *is-a*) was soon recognized as an important one, for which specialized built-in inference rules have been proposed: more specific nodes can be considered to "inherit" properties specified for the more general ones.

Semantic network systems differ from one another on a number of aspects. Some allow the "cancellation" of inherited properties accounting for exceptions to general rules; some have a single top node; some a single bottom node; some are complete lattices; some networks allow all link names to be specified by the user; others provide a set of predefined relations; some provide extended notations for the representation of the full first-order logic expressions (i.e. partitioned semantic networks (Fikes & Hendrix, 1977)).

In general, a semantic network system includes a set of structure manipulation primitives for adding and deleting nodes and links, and for traversing the graph. A few built-in inheritance mechanisms are provided; for different reasoning tasks, the users have to write their own search and inference procedures. The completeness and correctness of such implemented mechanisms is in general extremely hard to predict.

Semantic net notations are popular in language understanding programs and in contexts where the fundamental operation is the recognition of complex objects, for example in vision systems.

Frames

Another influential idea in Knowledge Representation has been that of frames introduced by Minsky (Minsky, 1981). While semantic networks emphasize small and primitive units of knowledge and their relations to one another, frames emphasize larger and complex chunks of knowledge. A frame is a cluster of information related to a concept including relations to other frames, attributes with their "typical" values and procedures.

The original conception of frames focused on their use as prototypical schemas against which to compare new situations with the goal of recognizing them, filling in the details, generating predictions or expectations.

The more concrete version of frames which has survived in representation languages is best described as a collection of slot/value pairs where the values may be among other things, defaults, demons (procedures which are activated on occurrence of specified events), and relations to other frames. Among frame-based representation languages are KRL (Bobrow & Winograd, 1977), FRL (Roberts & Goldstein, 1977), KEE (Fikes & Kehler, 1985), KRS (Steels, 1985).

Formal Accounts of AI Paradigms

The excessive freedom in the use of semantic network and frame formalisms, due to a lack of precise semantics, drew criticisms from a number of researchers. In particular Woods (Woods, 1975) and Brachman (Brachman, 1983) advocated the importance of a more formal account of the meaning of nodes and links in semantic networks. Hayes (Hayes, 1979) argues that the essential aspects of reasoning with frames can be captured by a first order formalization. Therefore frame based representations, at the representational level, are equivalent to a set of expressions in first order logic. Nevertheless some of the basic ideas such as inheritance and attributes have survived as useful structuring mechanisms, and are being given a more formal account in more recent knowledge representation systems. KRYPTON (Brachman, Gilbert Pigman & Levesque, 1985), KL-TWO (Vilain, 1985) and OMEGA (Attardi & Simi, 1981, 1986) are examples of systems in which logic is used to give formal semantics to the representation language and reasoning apparatus.

As a consequence of this formalization effort, properties of the representation system and of the inference machinery can be established, as for example, the possibility and complexity of deciding subsumption between concepts (whether one concept is more general than another one) (Brachman & Levesque, 1984) or a full axiomatization of the behaviour of inheritance and attributes in OMEGA (Attardi & Simi, 1981).

In general, concepts such as prototypes and defaults have been sacrificed in this effort towards formalization due to the fact that a satisfying formal treatment of the non-monotonic reasoning involved in using defaults is still under study (see section 4 for formal accounts of inheritance with exceptions). Moreover, unconditioned allowance for exceptions, claimed to be an important aspect of modelling knowledge of a natural kind, is in conflict with the possibility of automatic classification, an important reasoning method in all these systems. In fact the task of classification requires that criteria (sufficient conditions) for determining membership in a class be stated as part of the class definition, while most of the former frame-based representation languages allow only for necessary properties which can even be overridden in sub-classes (Brachman, 1985). As Brachman also points out, definitional capabilities (the ability to specify necessary and sufficient conditions), is also fundamental for defining new concepts in terms of more primitive ones (Brachman, 1985).

These considerations, together with the concern for the tractability of the reasoning task, have originated research on structural inheritance and terminological languages. We will review some of the proposals in the section about hybrid systems.

2.2   Logic Programming

As recalled in the introduction, pure LP consists of a resolution proof procedure (called SLD-resolution) which is able, given a program (i.e., a set of universally quantified formulas, called definite Horn clauses, of the form $A \leftarrow B_1 \wedge .... \wedge B_n$ with $n \geq 0$ where $A, B_1, ..., B_n$ are atoms) to check whether a formula (the goal) $\exists x_1 ... x_k. A_1 \wedge ... \wedge A_m$ is a logical consequence of the program.

The success of this new programming paradigm is due to Kowalski (Kowalski, 1974) who has given a procedural interpretation to this kind of theorem prover (clauses as procedure definitions, goal as the main program, resolution as procedure invocation, unification as parameter passing).

Another key observation has been that using this new paradigm, the user could concentrate more on the logical design of the program, leaving the control to the underlying language. This can be summarized in the formula Algorithm = Logic + Control (Kowalski, 1979b).

LP, thanks also to very efficient implementations, showed itself particularly well suited to express procedural knowledge, still maintaining the possibility of a clear logic semantics and the possibility to reason about programs.

Many systems based on procedural knowledge, first of all production systems, could be very well expressed using LP, by viewing rules as Horn clauses. As a matter of fact, PROLOG (the first and most popular LP language) has become a major AI implementation language, particularly used in expert systems.

KR formalisms are, in general, used in connection with the so called knowledge bases, which differ from traditional data bases in their deductive capabilities. A possible answer to the need of representing and manipulating a knowledge base, came from the research in LP and databases. In fact, the understanding that relational data bases could be easily seen as Horn clause programs, and then improved with the so called intentional part originated a new research field. An important book by Gallaire and Minker (Gallaire & Minker, 1978) introduced what are now called deductive data bases (Gallaire, Minker & Nicolas, 1984), which are closer to the idea of a knowledge base. The many proposals on how to efficiently integrate LP and data base technologies had a large influence both in expert systems and knowledge base techniques.

Moreover, Kowalski (Kowalski, 1979a), made clear that LP could influence more complex KR tasks. The key point was to show that a particular form of First Order Logic, clausal logic, could be used directly in many applications as a tool for problem solving and computer programming. The top-down behaviour of the goal directed inference, used in PROLOG, was also given a problem solving interpretation.

Many formal concepts such as clauses, Herbrand Universe (or universe of discourse) and Skolem constants were given a nice interpretation as useful tools to represent procedural and non-procedural knowledge as well as data bases and integrity constraints.

In (Kowalski, 1979a), it is shown that LP can be used to deal with some classical knowledge representation problems: semantic networks are represented by means of Horn clauses and the frame problem is given a solution by means of a clausal form axiom.

3. MULTIPLE PARADIGMS AND SPECIALIZED REASONERS

In both LP and KR, a key point in the newest approaches is the idea that cooperation and integration of different deductive mechanisms is a way to balance efficiency and expressiveness.

The trend towards hybrid systems in KR and the extensions and integrations of LP with other paradigms (functional and object-oriented) could greatly influence the design of an efficient and powerful KR Language.

3.1 Hybrid Systems

Each of the KR methods discussed in the previous section has its niche. However, in many cases, a complex problem calls for multiple types of reasoning competence. Hybrid systems originate from the consideration that it is usually more desirable to attempt to integrate a number of representation and reasoning paradigms rather than to force a representation to handle, in an awkward way, a task that another one handles smoothly.

Most of today's commercial environments for building knowledge based applications integrate different paradigms such as, most typically, a frame/object system and a rule system, semantic networks, and perhaps a logic programming component loosely integrated with the rest. Their major disadvantage is that they offer a number of tools, and correspondingly different languages and programming styles, with overlapping functionalities. Few indications are available to the user to decide which one to choose for developing an application.

A more interesting kind of hybrid system is that which resorts to specialized reasoning components to deal with the complexity of the reasoning task. Among these, good examples are Nelson and Oppen's work on cooperating decision procedures (Nelson & Oppen, 1979), Cake (Rich, 1985), which integrates a specialized planning language for reasoning about programs with a logical component and those systems which make the distinction between terminological and assertional knowledge, in particular Krypton (Brachman, Gilbert Pigman & Levesque, 1985) and Kl-Two (Vilain, 1985).

The separation between terminological and assertional knowledge has been advocated by the following motivation. Terminological knowledge has to do with the definition of terms (like in "A bachelor is a male adult who is not married") while assertional knowledge has to do with stating facts about the world by making use of those terms (like in "John is a bachelor"). Terminological languages require definitional capabilities and can benefit from an efficient kind of inheritance (*structural inheritance*) and classification. Moreover, the separation is also suggested for modularity reasons, since terminological knowledge is seen as more stable with respect to assertional knowledge, which is seen as contingent.

Krypton is a system which separates representation and reasoning into a terminological component (called the T-Box) and an assertional component (called the A-Box). The T-Box is a distilled version of the concept hierarchy of KL-One (Brachman & Schmolze, 1985). Two important kinds of relationship between concepts in the T-Box are subsumption and disjointness. The A-Box is essentially a resolution theorem prover for first order logic. Hybrid reasoning is achieved as follows: each unary and binary predicate in the A-Box is a pointer to a concept in the T-Box. During a resolution step in the A-Box, disjointness and subsumption information from the T-Box can be exploited to recognize the inconsistency of two literals.

KL-Two is the successor of the knowledge representation system KL-One (Brachman & Schmolze, 1985). Like Krypton, Kl-Two divides reasoning into a terminological and an assertional component. The terminological component, called NIKL, is again an evolution of

KL-One; the assertional component, called PENNI, is an implementation of RUP (Reasoning Utilities Package) (McAllister, 1982), which provides a data base for logical propositions, a truth maintenance system, unit propositional resolution and an incremental algorithm for maintaining congruence classes. Terminological information is exploited by PENNI, which performs only propositional deduction, by instantiating universal statements, corresponding to sub-class relations in NIKL. Communication among the two components goes also in the other direction; all the information about an individual are collected together and abstracted in a concept which is then classified in NIKL, thus extending the number of explicitly represented concepts.

The advantage of the terminological/assertional separation in these systems relies on reasoning about terminology being performed efficiently. This requires a careful computational analysis of the concept-forming operators (Levesque & Brachman, 1984). Research on terminological languages is therefore important in its own right.

Terminological languages formalize the basic ideas of frames. They include a set of syntactic constructs for concept definition in terms of other concepts, like for example conjunction and restriction on the attribute values and number of fillers. In this line of research the fundamental issue is to find the right compromise between expressivity of the language and tractability of the basic operations such as computing subsumption. Unfortunately it has been shown that algorithms of reasonable complexity are possible only for languages of very limited expressivity (Levesque & Brachman, 1984; Nebel, 1988).

The current proposals are very similar but differ in their position with respect to this trade-off. The T-box of KRYPTON has a very limited language and a complete and tractable subsumption algorithm. KL-One, NIKL, KANDOR (Patel-Schneider, 1984), opt for expressivity by giving up completeness of subsumption. Recent results have shown that subsumption is undecidable for NIKL and for terminological languages of the same power (Patel-Schneider, 1989b).

Patel-Schneider (Patel-Schneider, 1989a), instead of restricting the language, proposes a weaker semantics for terminological languages, based on a four-value logic (a truth value can be any subset of {true, false}). This non-standard semantics is able to account for a significant subset of the classical subsumption relation for which a complete and tractable proof procedure exists. The price to pay is a less intuitive semantics.

This concern for tractability has also originated the proposal to use, at least for the more immediate kind of reasoning that we seem capable of, more "vivid" kinds of representations, in direct correspondence with aspects of the worlds; this approach tries to reduce every piece of knowledge, even incomplete, to positive and concrete statements, perhaps making use of defaults, with the goal to make the complexity of queries similar to data base look-up (Levesque, 1986).

3.2   Extensions and Integration of LP with Other Paradigms

Constraint Logic Programming
The first natural extension of LP is to add sets of equations as programs, thus referring to first-order logic with equality. To limit the combinatorial explosion of the search space Plotkin (Plotkin, 1972) proposed to replace standard unification by semantic unification modulo the theory defined by the equational system. Semantic unification of two terms modulo a set of equations E corresponds to finding the substitutions that make the terms to belong to the same equivalence class. A semantic foundation of this approach can be found in (Jaffar, Lassez & Maher, 1984).

An evolution of this last result is the Constraint Logic Programming (CLP) Scheme (Jaffar & Lassez, 1987), a framework in which it is possible to give the formal foundations of a class of programming languages. CLP encapsulates both the paradigm of constraint solving and Logic Programming. One of the major advantages of the constraint solving paradigm is to allow implicit definition of properties; as an example consider the following explicit definition of a set of names

S = {Roberto Barbuti, Maurizio Martelli, Maria Simi,...}

which could be implicitly defined by the constraints

$S = \{x : salary(x) < 2300000 \ \& \ status(x) \neq full\_professor\}$.

The aim of adding constraints to Logic Programming is to replace the Standard Herbrand Universe by a user defined domain and unification by constraint solving. If some conditions on the constraint system are satisfied, the semantic properties of usual logic programming are preserved in CLP: in this sense, CLP is a scheme (Jaffar, Lassez & Maher, 1986). Many different instances of the scheme have been proposed and implemented.

CLP-like languages (CLP(R)) (Jaffar & Michaylov, 1987), CHIP (Dincbas, Van Hentenryck, Simonis, Aggoun, Graf & Berthier, 1988), ...) have a twofold impact on KR. On one hand they can easily accommodate specialized reasoners, such as real arithmetic problem solvers or any other kind of problem solver. On the other hand, considering equations (and/or disequations) as part of the language makes LP a richer language in expressive power.

Logic-functional integration

The future of Logic Programming as a powerful programming language (with a great expressive power, but also very efficient) has to consider the important achievements of different programming paradigms and in particular functional and object oriented programming.

Logic and Functional Programming are the two most popular styles of AI programming. The debate on which paradigm is best suited for AI has been substituted by discussions on how to combine the two paradigms, and thus the advantages of both (without their drawbacks).

Functional languages share with logic languages the formal simplicity (function application as the only control construct) and the property of being based on a well-established mathematical theory, in this case (some kind of) λ-calculus, which directly provides a clear semantics. Reduction is the key concept, which corresponds to the use of equalities as rewrite rules, in contrast with standard logic programming, where the equality is not explicitly present.

Apart from notational aspects, the fundamental difference with the languages based on first-order logic is the presence of the notion of higher-order function, a powerful structuring concept which can be exploited in programming in the large, in program synthesis from specifications, etc.

In addition, functional languages offer a variety of other programming concepts, such as typing disciplines, outermost/innermost strategies, etc., which have proved very useful in many AI applications, and would be profitably included in an integrated system.

Several different approaches to the integration have been proposed (see (DeGroot & Lindstrom, 1986) where many approaches are presented). Roughly, they can be partitioned into two classes. The first class is the logic+functional languages, i.e. the logic languages enhanced with some limited functional features, essentially based on first-order logic with equality: they usually lack one of the very aspects which characterize functional programming, i.e. higher-order capabilities (Barbuti, Bellia, Levi & Martelli, 1985, 1986; Bosco, Giovannetti, Levi, Moiso & Palamidessi, 1987; Goguen & Meseguer, 1984). The second one is the functional+logic languages, i.e. the functional languages augmented with logic capabilities, for example the ability to solve equations, to invert functions, etc. (term rewriting systems and equational programming influenced these proposals, see (Klop, 1987;

O'Donnell, 1985)). For a survey of the most relevant proposals of both kinds see (Bellia & Levi, 1986).

If the functional part is constrained to be first-order, the integration can be achieved in a purely logical framework, namely Horn clause logic with equality. The language Kernel-LEAF (Bosco, Giovannetti, Levi, Moiso & Palamidessi, 1987), an evolution and refinement of the language LEAF (Barbuti, Bellia, Levi & Martelli, 1986) follows this approach and has a functional component (defined by equations) with an operational semantics based on unification and which exhibits the same properties of standard logic programs (logical variables, search, function invertibility).

Many integration proposals base the operational semantics of the functional component on narrowing (Dershowitz & Plaisted, 1985; Fribourg, 1985); K-LEAF is based on the transformation of equations to *flat form* (where function composition is eliminated and replaced by the logical operator AND) and on the SLD-resolution inference rule.

Obviously these kinds of extensions of the LP paradigm must be based on a formal semantics (model theoretic and fixed point) to maintain the nice characteristics of LP, while gaining in representation power and efficiency.

Types

The problem of types in a new integrated programming paradigm is also crucial, and it could be thought as the counterpart of the terminological component of many KR formalisms discussed above.

There have been many proposals to introduce types in LP (see as an example (Mycroft & OKeefe, 1984)). Besides the classical advantages of types, this information can also be used to reduce the search space by extending the basic unification mechanism to "order-sorted algebras" (Meseguer, Goguen & Smolka, 1980).

Types are also very important to study how to integrate functional and Object Oriented programming, and thus understand better the possibility of a well founded integration of all three paradigms (Goguen & Meseguer, 1987; Ait-Kaci & Smolka, 1987). Subtyping or type inclusion are very much related to the concept of inheritance.

The LOGIN language (Ait-Kaci & Nasr, 1986) extends Prolog to accommodate taxonomic ordering relations between constructor symbols, thus obtaining the separation of inheritance from the rest of the logical inference machinery.

A very interesting generalization of this approach comes from the theorem proving field: Theory Resolution (Stickel, 1985). Theory resolution incorporates specialized reasoning procedures in a resolution theorem prover by resolving on sets of literals whose conjunction is determined to be unsatisfiable in a specialized theory, rather than by just using ordinary unification of complementary literals. As seen above, this seems a key idea both in AI systems as in many extensions of LP and Theory Resolution, given its general nature, is probably a theoretical framework able to encompass many proposals bridging the gap between LP and KR.

Modularization and Integration with Object-oriented Languages
LP for KR lacks features for structuring large amounts of dynamically evolving knowledge. Object Oriented programming is particularly suited for this purpose. Thus the study of possible integrations of these paradigms is of vital importance. The need for modularization in LP has been recognized very early in the field.

Many proposals for the introduction of modules in logic programming have been presented (Fitting, 1987; OKeefe, 1985; Mancarella, Pedreschi & Turini, 1988) which are characterized by two design criteria:

The modularization mechanism allows the building of programs out of pieces; and
There exists a clear semantics based on that of logic programs.

In (Miller, 1986) an approach to modularization, based on intuitionistic implication, is presented. The main idea is to allow the implication connective in goals and clause bodies in order to model the passing of predicate definitions from one module to another. This method requires higher-order features. The paper shows how to build models for these programs by a fixed point construction and it presents an operational semantics sound and complete for intuitionistic logic.

Intuitionistic implication, together with classical implication, is also used in (Giordano, Martelli & Rossi, 1988) in order to have structuring mechanisms with static scope rules in logic programs.

In (Nait Abdallah, 1986) a theory of modules based on second order logic is presented. A *procedure definition / procedure call* mechanism and algebraic structures are defined as conditional procedures. This is obtained by tying together the first-order predicate logic approach and the λ-calculus approach. The main goal of this construction is to introduce the notions of *hierarchization* and *modularization* into the usual framework of logic programming.

In Eqlog (Goguen & Meseguer, 1984) types and modules of algebraic languages (Obj, Clear) are introduced in a logic language. In this context, the arguments of modules are theories. Therefore, formal parameters express theories which must be subsumed by the actual parameters. This means that each axiom of the formal parameter theory must be a theorem of the actual one.

Note that since modules can define several types, when applying a generic module to an argument, it may be necessary to specify a correspondence between types, functions and predicates of the actual one. This can be done by creating a particular view of a module and then using this view as an actual parameter.

The modularization approach of Eqlog is extended, in (Goguen, 1986), to a more general setting. The proposed concept of *institutions* formalizes the logical systems suitable for a programming methodology and seems useful to unify programming paradigms. Institutions allow the modelling of programming and specification language characteristics from a more abstract level. In particular, first-order logic, Horn clause logic, equational logic, etc. are institutions. As a consequence, the module mechanism of Clear (extended to institutions) allows the composition of pieces of programs written in different logic programming languages, and, obviously, the provision of a modularization mechanism for these languages.

The problem of a deeper integration between LP and Object Oriented programming, that takes into consideration also the problem of dynamic changing knowledge, has received less attention than the integration of LP and functional programming or the problem of modularization. Most of the proposals are essentially implementations of some kind of objects in LP.

McCabe, in (McCabe, 1988), proposes the notion of an object as a labeled theory and models some distinctive features of OOP by translating them into Horn Clauses. The problem of objects with states is not yet considered. Similar approaches are in (Stabler, 1986; Zaniolo, 1984). The main features of Gallaire's system POL (Gallaire, 1986) are backtrackable method calls with a success/failure semantics, variable objects calls, and default and deterministic method evaluation mechanisms. In this proposal we can see the convergence of results in AI hybrid systems, semantic data base systems, theorem provers using theory resolution (Stickel, 1985) and extended LP paradigms (as higher order).

Other approaches are those that see objects as perpetual processes where the state is represented by the current goal list (some literals representing specific objects) (Shapiro & Takeuchi, 1983; Kahn, Tribble, Miller & Bobrow, 1986). Formal semantics for these approaches inherit the problems of giving semantics to concurrent languages.

There is also a proposal (Chen & Warren, 1988) which bases the integration of LP and OOP on the intentional logic of Montague. A new type of variable (intentional) is introduced into first order logic to represent objects whose meanings are functions from states to values. Correspondingly new types of predicates and clauses (dynamic ones) are introduced to deal with them. The semantics is given in terms of enriched semantic structures which take into account the new concepts.

The integration of the foundations of logic, functional, and object oriented programming is considered a very important step to be able to build a really powerful programming language that is not simply the union of different types of programming paradigms but a coherent and semantically clear language. Some important contributions in this direction are (Gallaire, 1986; Goguen & Meseguer, 1987; Ait-Kaci & Smolka, 1987; Ait-Kaci & Lincoln, 1988).

4.   NON-MONOTONIC REASONING

Many reasoning tasks require reasoning with incomplete information by making assumptions based on typical behaviour and other common sense knowledge. This is a kind of non-deductive activity which presents a challenge for reasoning systems based in any way on standard logic. Standard logic is monotonic, once a conclusion has been reached new information can never invalidate it but much of human reasoning is non-monotonic, in that new information can invalidate old assumptions.

One of the most active research areas in knowledge representation is the study of non-standard logics to account formally for systems able to perform semantically coherent non-monotonic reasoning.

Non-monotonic reasoning very soon became an important topic in LP also. In fact, Horn clauses do not allow negative literals in their bodies, and already in (Clark, 1978) the need for them was recognized. The problem was to reach a compromise between the full power of general clauses and the efficiency of SLD-resolution. This compromise was a new inference rule called Negation as Failure (and correspondingly SLDNF-resolution). Recent research in negation in LP has also shown strong relations with many results in non-monotonic logic.

4.1 Logics for Non-monotonic Reasoning

There are essentially two types of approaches to non-monotonic formal reasoning:

Completion approaches: where the goal is to get a complete theory (a theory such that every ground atom in the language or its negation are in the theory) or some weaker form of completeness, i.e. with respect to a predicate or set of predicates. The Closed World Assumption, predicate completion and circumscription are in this class.

Non-standard logics: that is logics with special inference rules (Reiter's default logic (Reiter, 1980), non-monotonic logics (McDermott & Doyle, 1980; McDermott, 1982)) or a different proof theoretic characterization of the set(s) of formulas non-monotonically derivable from a set of assumptions (for example the Autoepistemic Logic (Moore, 1983)).

The Closed World Assumption, due to Reiter (Reiter, 1978), produces a complete theory according to the following rule: "if you cannot prove a ground formula α in the language, then assume $\sim\alpha$". This assumption has been extensively used in data base applications.

In both Predicate Completion and Circumscription the goal is to find a formula f, called *completion formula,* that added to a set of beliefs S restricts the set of objects which satisfy S to only those which have to satisfy S (Genesereth & Nilsson, 1987).

Predicate completion does so with respect to a predicate, or set of predicates. For example if the set of beliefs contains only P(A), since this is equivalent to $\supset x\ (x=A) \supset P(x)$, the completion formula is $\supseteq x\ P(x) \supset (x=A)$, stating that the only object which satisfies P is A. Predicate completion is defined and proved consistent for theories in a restricted, syntactically characterized, class (the class of clauses *solitary* in a predicate), in which all the formulas involving a predicate P can be reduced to the general form $\supset x\ E \supset P(x)$, where E is an existentially quantified formula not involving P. In this case the completion formula is $\supseteq x\ P(x) \supset E$ (Clark, 1978; Genesereth & Nilsson, 1987).

The definition of circumscription (McCarthy, 1980), is based on the model theoretic notion of *minimal model*. Essentially, a model M is minimal with respect to a predicate P, if the set of objects which satisfy P in M is a subset of the set of objects which satisfy P in any other model of the theory. The general formula which is added to a set of beliefs S to achieve the circumscription of a predicate P is the following:

$$\supset P^*\ S(P^*) \wedge ((\supseteq x\ P^*(x) \supset P(x)) \supset (\supset x\ P(x) \supset P^*(x)))$$

stating that there can be no predicates, satisfying the set of beliefs S, with a smaller extension than P. This is a completely general mechanism; the only problem is that this second order formula is practically useful only in special cases, for example when it can be proved to collapse to a first order formula: results in this sense are mainly due to Lifschitz (Lifschitz, 1985). It has been shown that predicate completion is a special case of circumscription (Reiter, 1982).

Default Logic
In Reiter's default logic (Reiter, 1980) the problem of non-monotonic reasoning is dealt with by using ad hoc inference rules. A default rule is

an inference rule used to augment a theory under certain restrictions. The default rule

$$\frac{\alpha : \beta}{\gamma}$$

can be informally interpreted as follows: "if α can be proved, and β can be consistently assumed, then γ can be assumed".

An extension of a default theory, composed of a set of assumptions W and a set of default rules D, has the following properties:

1. It is closed under deduction;
2. It includes W;
3. For any default $\alpha{:}\beta/\gamma$, if α is in the extension and $\sim\beta$ is not in the extension, then γ is in the extension.

Extensions are guaranteed to exist for normal default theories (theories where the defaults have the form $\alpha{:}\beta/\beta$). In the case of interacting defaults multiple extensions are possible.

Reiter shows that default logic is not even semidecidable but describes a resolution theorem prover that can be used for top-down or backward searches for default proofs (Reiter, 1980).

Non-monotonic Logics

One of the first attempts to develop a logic for non-monotonic reasoning was made by McDermott and Doyle (McDermott & Doyle, 1980). They presented a version of the predicate calculus with an operator "M", to be informally interpreted as "it is consistent", and an additional inference rule "If it not possible to *infer* $\sim\alpha$ then $M\alpha$", where *infer* includes the use of the inference rule itself. A second version of this logic was presented in (McDermott, 1982) taking different versions of modal logics as a starting point rather than predicate calculus. Several undesirable properties of this logic were pointed out (Moore, 1983), among which were the lack of an intuitive semantics and of a formal proof of consistency for the version with quantifiers.

Autoepistemic Logic

Autoepistemic logic (Moore, 1983) is an interesting contribution to the field. Autoepistemic reasoning is intended to model the reasoning of an ideally rational agent reflecting upon his beliefs. The modal operator L is used with the meaning "is believed". An autoepistemic theory T, should have the following properties:

1. It should be closed under deduction;
2. If α is in T, then $L\alpha$ is in T;
3. If α is not in T, then $\sim L\alpha$ is in T.

Such theories, called stable, are complete with respect to what is believed, in the sense that they contain any sentence that the agent is justified in believing. Moreover soundness of the agent's beliefs with respect to a set of premises S is guaranteed by the syntactic notion of theory *grounded in* S (every sentence of the theory T is included in the tautological consequences of S ∪ {Lα: α is in T} ∪ {~Lα: α is not in T}). A stable autoepistemic theory grounded in A, called stable expansion of A, is therefore a good characterization of the set of beliefs of an agent holding the premises A.

Autoepistemic theories provide a rational reconstruction of the non-monotonic logics of McDermott (McDermott, 1982), which is seen as a logic of belief, explaining some of the peculiarities of that logic.

Hierarchic autoepistemic logic is a variant of autoepistemic logic in which the belief set is partitioned in subtheories hierarchically organized. This proposal is motivated by the utility to include in the logic, preference criteria among defaults; in addition it is possible to give a constructive definition for the autoepistemic operator in contrast to the usual self-referential fixed-points. The result is a theory easier to implement (Konolige, 1988).

A lot of work is being done in comparing different formalisms for non-monotonic reasoning, for example Konolige analyzes the connections among default theories and autoepistemic logic (Konolige, 1987).

Accounting for Inheritance with Exceptions
Inheritance systems (either based on frame or semantic nets) exhibit a certain non-monotonic behaviour, by allowing for explicit exceptions or overriding of properties in sub-classes; inheritance is computed efficiently by searching the inheritance graph from specific to general information, according to the intuition that the most specific information wins. However, in the absence of adequate semantic characterization, the inference in these systems was only defined through intuitions and the actual behaviour of the programs. Especially in the case of multiple inheritance and parallel algorithms the behaviour of the systems was often unpredictable.

NETL is, for example, an early system which exploits a technique of parallel marker propagation for inheritance reasoning. The correctness of the algorithms used by NETL was left to one's intuition (Fahlman, 1979).

A first account of the semantics of inheritance in NETL-like systems with explicit exceptions was given in terms of Reiter's default logic (Etherington & Reiter, 1983). This is done by establishing a correspondence between inheritance hierarchies and default theories, and between sets of correct inferences and extensions of the default theory. Multiple extensions of a default theory come to correspond to multiple inheritance. A shortcoming of this approach is that it does not support the intuition, which is also a well established practice in inheritance systems, that "subclasses override superclasses", but rather is able to deal only with explicit exceptions.

In order to capture that intuition, Touretsky proposes a formal model for inheritance systems based on a principle called *inferential distance ordering*: essentially "α is closer to β than γ, if there is an inference path from α to β through γ". The superclass from which a subclass should inherit is determined according to this ordering (Touretzky, 1986). Touretsky also discusses how the inferential distance ordering idea can be applied to default logic by exploiting an order of application on default rules induced by the inheritance graph (Touretzky, 1986).

Sandewall analyzes multiple inheritance systems with exceptions, in terms of a number of structure types, claiming that combinations of those structure types can be taken as a partial semantics on inheritance systems (Sandewall, 1986).

Finally formal accounts of inheritance in terms of predicate *completion* and circumscription are presented in (Brewka, 1987; Genesereth & Nilsson, 1987).

4.2   Logic Programming and Computational Negation

Negation as Failure

Negation as failure is the usual inference procedure for negative information in logic programming. It can be considered as the computable counterpart of the Closed World Assumption of Reiter (Reiter, 1978).

The Closed World Assumption states that, given a first order theory T, it is possible to infer the negation of a formula ~A, if A is not provable from T. Unfortunately this is not an effective inference rule because of the semi-decidability of the inference process in first order logic.

Negation as failure (Apt, 1987; Clark, 1978; Kunen, 1987; Shepherdson, 1984, 1985, 1987) allows the inference ~α *iff* the proof of α fails in a finite amount of time; for this reason it is also called computational negation. The reference theory for this inference rule is not the program, but a theory (called completion) containing the program with the *iff* connectives replacing the *if* ones, plus some axioms for the equality.

Although the correctness and completeness of this new inference rule with respect to the completion has been proved (Jaffar, Lassez & Lloyd,

1983), the use of negation as failure in a SLD-Resolution based system leads to incompleteness (Lloyd, 1987; Barbuti & Martelli, 1986; Kunen, 1989). In fact, when the program clauses with negative literals in their bodies (called *general* or *normal*) are considered, the negation as failure rule has to be incorporated into the proof procedure (called SLDNF-resolution). In many cases, this proof procedure is not complete with respect to the program completion. Furthermore, the completion of a general logic program could be inconsistent.

An attempt to restrict the class of general logic programs in order to have useful properties is given in (Apt, Blair & Walker, 1987); a particular class of programs is defined (*stratified*) whose completion is always consistent. These programs rule out recursion through negated literals in the body of the clauses. However, when using computational negation, there are still stratified programs for which the unique minimal model property is lost. This motivates the need to find a particular model (usually a minimal one) which is considered as the semantics of a logic program (Lifschitz, 1986; Przymusinski, 1986). In particular, in (Przymusinski, 1986) a semantics based on perfect models is defined. Roughly speaking, a minimal model M is perfect if the set of atoms whose predicate symbols occur in negative form in clause bodies is minimal, with respect to other minimal models.

Negation as failure is a non-monotonic operator, in fact, extending the theory by new clauses could result in a smaller set of theorems. The non-monotonic character of this operator, closely relates logic programming to non-monotonic reasoning. As a matter of fact, it can be shown that the use of negation as a failure rule can implement some of the approaches presented in the previous subsection.

In (Przymusinski, 1990) the semantics of logic programs by means of perfect models is considered. It is shown that, for stratified programs, prioritized circumscription, default logic and autoepistemic logic are equivalent to the perfect model approach.

Computational Negation

A transformational approach to computational negation (Barbuti, Mancarella, Pedreschi & Turini, 1987, 1990), called *intensional negation*, represents, in the field of logic programming, the current trend in vivid forms of knowledge (Levesque, 1986).

This method, based on a representation of the complement of terms (Lassez & Marriot, 1987), is able to derive from a program, a new program in which the negative information is intensionally represented. The negation of a predicate P is represented by a new predicate P~ whose solutions are the ones provable by the negation as failure of P. The difference is that P~ is able to generate solutions and not only to check for the validity of a ground negative literal.

The transformation from a logic program with intensional negation to a "vivid" data base is straightforward. Moreover, this approach allows one to solve one of the main drawbacks of negation as failure, i.e., the applicability to only ground negative literals.

5. REASONING ABOUT KNOWLEDGE AND BELIEF

Another important problem in KR is the ability to represent and reason about other agent's propositional attitudes such as knowledge and beliefs, but similar problems also arise for truth and provability, goals, intentions and so on. We will describe essentially two approaches which come from KR. The first one is the meta-level approach which is also becoming much used in LP; common problems in manipulation of programs and calls to the manipulated programs need a meta level approach and suitable language primitives. Many projects are underway to incorporate it in an efficient way into LP systems.

5.1   Logics for Knowledge and Belief

In representing propositional attitudes such as "knowledge" and "belief", the key representational problem is that they give rise to referentially *opaque* contexts, namely it is not possible to substitute equivalent expressions inside the corresponding operators. A classical example is that the fact "Bel(John, phone(Bill)=277-1265)" together with   "phone(Katy)=277-1265"   should   not   imply   that "Bel(John,phone(Bill)=phone(Katy)", that is the ordinary logical rule of substitutivity of equals should not apply in the context of these operators.

There are essentially two approaches to this problem:

Syntactic first-order approaches, in which Bel and Know are ordinary predicate symbols and their arguments are first-order terms chosen to represent a sentence;

Modal approaches, in which Bel and Know are modal operators and their arguments first order formulas.

Syntactic First-order Approaches

In syntactic first-order approaches the substitutivity of equals is naturally blocked by using names of formulas rather than formulas themselves. In addition to a suitable naming mechanism, some sort of linking rule connecting the names of formulas with their truth theoretic values in the theory is necessary; without it, the domain of syntactic objects remains completely disjoint from the domain of discourse in the real world. Examples of these linking rules are True("α") <–> α (Perlis, 1985), the reflection principle in FOL (Weyhrauch, 1980) and similar reflection rules in (Bowen & Kowalski, 1982; Attardi & Simi, 1984; Simi & Motta, 1988), connecting a meta-level notion of provability to the existence of proof in the object theory.

These linking rules have to be carefully formulated to avoid the introduction of inconsistencies in the theory. Early negative results in this sense, due to Montague (Montague, 1963), have diverted the research away from first-order approaches towards modal logics. But, as pointed out by Perlis (Perlis, 1988), Montague's negative results are due to "fundamental expressive strengths of first-order languages" with respect to modal ones. And in fact weaker first-order analogues of modal theories could be proved consistent (des Rivieres & Levesque, 1986). Perlis argues that, in particular, self-referential capabilities are necessary (for representing notions such has "g has a false belief, this itself being a belief of g") and are missing in modal logic. In (Perlis, 1985), Perlis proposes a consistent self-referential theory by weakening the linking rule True("α") <–> α.

Based on the meta-level approach, architectures which include a metalevel have been proposed in which it is possible to reason about syntactical aspects such as the form of propositions, proofs, and theories. In these systems the set of beliefs of an agent is modelled as provability with respect to a theory (an agent believes a iff a can be derived from the set of assumptions of the agent). The FOL system of Weyrauch is the first system with a general metalevel architecture (Weyhrauch, 1980), but see also (Aiello, Nardi & Schaerf, 1988; Bowen & Kowalski, 1982; Attardi & Simi, 1984; Simi & Motta, 1988) for applications of this idea and different proposals.

Modal Logics and Multiple World Semantics
The idea of "possible world semantics" for modal logics, first formalized by Hintikka (Hintikka, 1962), is that in each state of the world, an agent has a number of worlds that he considers possible (accessible from his world). An agent "knows" α exactly if α is true in all the worlds that he considers possible. By imposing conditions on the accessibility relation one can support a number of interesting properties of knowledge.

Given the flexibility of possible world semantics, this has been the basis for the development of various modal logics for knowledge and belief (see (Halpern & Moses, 1985) for a careful analysis). For a realistic model of human reasoning though, extension or modifications of this approach are necessary to deal with the problem of "logical omniscience"; in fact with possible world semantics if an agent knows p, he also knows all the logical consequences of p. This is not an acceptable model of human reasoning for several reasons: lack of awareness of the concepts; limited computational resources; ignorance of relevant rules; and locality of reasoning. Several proposals try to address these aspects (Levesque, 1984; Fagin & Halpern, 1985).

Other important developments, not discussed here, are concerned: with reasoning in a multiple agent context; about knowledge and action (Moore, 1985); about other mental attitudes such as goals and intentions; and time and change (see (Kowalski & Sergot, 1986) for an approach based on negation as failure).

5.2   Meta-level in Logic Programming

There are some proposals to extend logic programming with the possibility of reasoning in different theories and with different inference rules (Kauffmann & Grumbach, 1986a, 1986b; Furukawa, Takeuchi, Kunifuji, Yasukawa, Ohki & Ueda, 1984).

Prolog/KR (Nakashima, 1984) has been one of the first attempt to combine frame theory and logic by extending Prolog. It has multiple definition spaces, called multiple worlds, which can be combined dynamically on demand. Each world represents some concept and they are combined to form a conceptual hierarchy. Multiple inheritance and dynamic change of the viewpoint can be handled.

An interesting approach is the one in (Coscia, Franceschi, Levi, Sardu & Torre, 1988a, 1988b), which is based on meta programming. A formula can be proved with respect to different theories. Each theory has a specific inference procedure. The different inference rules are simulated by meta-interpreters.

The introduction of a meta-level in logic programming was firstly presented in a paper by Bowen and Kowalski (Bowen & Kowalski, 1982). This paper represented the starting point for various research about the extension of logic programming with reflection capabilities and about the use of such a new language(Bowen & Weinberg, 1985; Mancarella, Pedreschi & Turini, 1988; Sterling, 1985, 1988; Sterling & Beer, 1986; Safra & Shapiro, 1986; Bacha, 1987).

The amalgamation of the object language and the meta-level requires a method to represent the syntactic entities of the object language as terms of the meta-language and to implement an interpreter (meta-interpreter) for such terms in the meta-level.

The expressiveness of the amalgamated system is increased with respect to the object level mainly for two reasons (Coscia, Franceschi, Levi, Sardu & Torre, 1988a):

1. It allows one to express formulas which combine object level relations with meta-level relations;
2. The meta-interpreter can be extended with new inference rules, thus supporting new language features.

An advantage in using meta-interpreters in logic programming is given by the use of *partial evaluation* techniques (Ershov, 1977; Levi & Sardu, 1988; Gallagher, 1986; Takeuchi & Furukawa, 1986; Kursawe, 1986). Partial evaluation is a transformation technique that, given a program P, and a goal G, results in a new program P^1 which is a specialized version of P for goals with the structure of G. For such goals, in most cases, P^1 is computationally more efficient than P.

If P contains a meta-interpreter, the partial evaluation can result in a program P^1 which is an object level program. In this case the use of meta-level programming and partial evaluation combines the expressive power of a meta-level architecture with the efficiency of the object level.

6.   HIGHER ORDER APPROACHES

Many formalisms in KR and many extensions in LP either explicitly or for the purpose of giving a clear semantics use some kind of higher order logic (modal logics can easily be accommodated in this framework).

The idea to have, at least for semantical reasons, a unifying framework is very attractive, but recently something more has been proposed, namely some real programming languages or formalisms based explicitly on some higher order notions (see (Church, 1940) for the basis of the simple theory of types, and (Andrews, 1972) for a work that influenced very much the following proposals).

In LP the most popular is λ–Prolog (Nadathur & Miller, 1988), a logic programming language that extends Prolog by incorporating notions of higher order functions, λ-terms, higher order unification (Huet, 1975), polymorphic types, and mechanisms for building modules and secure abstract data types (see also (Bosco & Giovannetti, 1986) for another proposal). The theoretical foundations are given by the intuitionistic higher order theory of hereditary Harrop formulas. These formulas have the property to maintain some proof theoretic, nice characteristics of the Horn clauses; they can be thought of as the higher order correspondents of Horn clauses. These characteristics are well described with the notion of uniform proof (see (Miller, Nadathur & Scedrov, 1987) for more details).

Let us mention also another nice application of these concepts, namely the possibility to use a higher order logic language to specify theorem provers (Felty & Miller, 1988). These languages naturally specify inference rules for various inference systems. Higher order unification, which provides sophisticated pattern matching on formulas and proofs, can be used to determine when and at what instance an inference rule can be employed in a search for a proof. Tactics and tacticals can also be directly implemented (see also (Gordon, Milner & Wadsworth, 1979; Constable, 1986; Paulson, 1986) for related theorem provers).

The use of higher order logic is also a key point for the proposals of Nait Abdallah (Nait Abdallah, 1986, 1987). In a recent paper (Nait Abdallah, 1988) he shows some interesting relations between his approach, called Heuristic Logic, and many problems present in Learning, especially by formalizing different kinds of learning such as inductivism, dogmatic and sophisticated falsificationism.

There are many other possibilities to unify, at least at the semantics level, what is necessary for a good KR language. Institutions (Goguen & Burstall, 1985) are the basis for the attempt to integrate logic functional and object oriented programming in FOOPlog (Goguen & Meseguer, 1987).

Categories (Cousineau, Curien & Mauny, 1985) and the theory of Constructions (Coquand & Huet, 1985) are other very promising approaches that would require much more space to discuss. Let us only mention that recently there have been attempts (see (Lambek & Scott, 1986) for a good introduction) to reconcile mathematical logic and category theory. This has been obtained by showing the strong relations between typed λ-calculus, Martin-Löf type theory, (usual) intuitionistic type theory, and (respectively) cartesian closed categories, locally cartesian closed categories and toposes.

Such results strengthen our conviction that it is possible and necessary to work in the direction of the unification of various knowledge representation formalisms. This does not mean to have a unique formalism, but to have as many formalisms as necessary, with (possibly) their own efficient proof procedures, that are unifiable at the semantic level.

7. CONCLUSION

In the paper we have shown how some important aspects of the research in KR are being taken into account by the newer developments in LP.

The use of LP to solve KR problems is a very promising research direction and the cross-fertilization between the two fields can be very valuable. The role of the European research is particularly important in this area as LP has a very strong tradition in Europe and many of the best research groups are European. These facts contributed to the starting of significant projects within the ESPRIT Basic Research Action, which specifically addresses some of the themes discussed in this paper. Among them are the project Integration (CWI, Imperial College, LIENS, Philips, Lisbon University and Pisa University) which has the objective to integrate the foundations of Logic, Functional and Object Oriented programming, and the project Computational Logic (Imperial College, ADER, ECRC, and the Universities of Bristol, Edinburgh, Kaiserslautern, Leuven, Lisbon, Marseille, Passau, Pisa, Roma, Tübingen, Uppsala) whose goal is to develop the foundations for an integrated, logic-based software environment for knowledge-rich applications.

REFERENCES

Ait-Kaci H., Nasr R., "LOGIN: A Logic Programming Language with Built-in Inheritance", J. of Logic Programming, 3 (3), 1986, 187-215.

Ait-Kaci H., Smolka G., "Inheritance Hierarchies: Semantics and Unification", MCC Tech. Rep. AI-057-87, 1987.

Ait-Kaci H., Lincoln P., "LIFE: A Natural Language for Natural Language", MCC Tech. Rep. ACA-ST-074-88, 1988.

Aiello L., Nardi D., Schaerf M., "Reasoning about Knowledge and Ignorance", Proc. Int'l Conf. on Fifth Generation Computer Systems, 1988, 618-627.

Andrews P.B., "Resolution in Type Theory", Journal of Symbolic Logic 37, 1972, 395-397.

Apt K.R., "Introduction to Logic Programming", Tech Report CS-R8741, CWI, Amsterdam, 1987.

Apt K.R., Blair H.A., Walker A., "Towards a Theory of Declarative Knowledge", in Foundations of Deductive Databases and Logic Programming (J. Minker Ed.), Morgan Kaufman, Los Altos, 1987.

Apt K.R., van Emden M.H., "Contributions to the Theory of Logic Programming", J. ACM, 29, 3, 1982, 841862.

Attardi G., Simi M., "Consistency and Completeness of Omega, a Logic for Knowledge Representation", Proc. of 7th International Joint Conference on Artificial Intelligence, Vancouver, 1981.

Attardi G., Simi M., "Metalevel and Reasoning across Viewpoints", Proc. of ECAI, Pisa, 1984.

Attardi G., Simi M., "A Description Oriented Logic for Building Knowledge Bases", Proceedings of the IEEE, Vol 74, No 10, October 1986, 1297-1472.

Bacha H., "Meta-level programming: a compiled approach", in Logic Programming (Jean-Louis Lassez, Ed.), Proc. of the Fourth Intl Conf. on Logic Programming, The MIT Press, 1987, 394-410.

Barbuti R., Bellia M., Levi G., Martelli M., "On the integration of logic programming and functional programming", Proc. 1984 Symp. on Logic Programming, IEEE Comp. Society Press, 1985, 160-166.

Barbuti R., Bellia M., Levi G., Martelli M., "LEAF: A language which integrates logic, equations and functions", in (DeGroot & Lindstrom, 1986), 201-238.

Barbuti R., Mancarella P., Pedreschi D., Turini F., "Intensional Negation of Logic Programs: examples and implementation techniques", Proc. of Int. Joint Conf. on Theory and Practice of Software Development, TAPSOFT 87, LNCS 250, 1987, 96110.

Barbuti R., Mancarella P., Pedreschi D., Turini F., "A transformational approach to negation in logic programming", in Journal of Logic Programming, 8 (3), 1990, 201-208.

Barbuti R., Martelli M., "Completeness of the SLDNF-resolution for a Class of Logic Programs", Third Int. Conf. on Logic Programming, London, 1986, 600-614.

Bellia M., Levi G., "The relation between logic and functional languages: A survey", Journal of Logic Programming 3, 1986, 217-236 .

Bobrow D.J., Winograd T., "An Overview of KRL, A Knowledge Representation Language", Cognitive Science 1 (1), 1977, 3-46, reprinted in (Brachman & Levesque, 1985).

Bosco P.G., Giovannetti E., "IDEAL: An Ideal Deductive Applicative Language", Proc. 1986 Symp. on Logic Programming, IEEE Comp. Society Press, 1986, 89-94.

Bosco P.G., Giovannetti E., Levi G., Moiso C., Palamidessi C., "A complete semantic characterization of K-LEAF, a logic language with partial functions", Fourth Symposium on Logic Programming, San Francisco, 1987.

Bowen K.A., Kowalski R.A., "Amalgamating Language and Metalanguage In Logic programming", in Logic Programming (K. Clark and S. Tarnlund eds), Academic Press, 1982, 153-172.

Bowen K.A., Weinberg T., "A Metalevel Extension of Prolog", Proc. Symp. on Logic Programming, Boston, 1985, 48-53.

Brachman R.J., "What IS-A Is and Isn't: An Analysis of Taxonomic Links in Semantic Networks", Computer IEEE, 1983.

Brachman R.J., "I lied about the Trees: or Defaults and Definition in Knowledge Representation", Ai-Magazine, Fall 1985.

Brachman R.J., Gilbert Pigman V., Levesque H.J., "An Essential Hybrid Reasoning System: Knowledge and Symbol Level Accounts of Krypton", Proc. of 9th IJCAI, Los Angeles, 1985.

Brachman R.J., Levesque H.J., "The Tractability of Subsumption in Frame Based Description Languages", Proc. of AAAI-84, Austin, TX, August, 1984.

Brachman R.J., Levesque H.J., Readings in Knowledge Representation. Morgan Kaufmann, 1985.

Brachman R.J., Schmolze, "An Overview of the KL-ONE Knowledge Representation System", Cognitive Science 9(2), April-June, 1985.

Brewka G., "The Logic of Inheritance in Frame Systems", Proc. of IJCAI-87, 1987, 483-488.

Bundy A., "A Broader Interpretation of Logic in Logic Programming", Proc. of Fifth Symposium and Conference on Logic Programming, Seattle, 1988, 1624-1648.

Chen W., Warren D.S., "Objects as Intentions", Proc. of Fifth Symposium and Conference on Logic Programming, Seattle, 1988, 404-419.

Church A., "A formulation of the simple Theory of Types", Journal of Symbolic Logic 5, 1940, pp. 56-68.

Clark K.L., "Negation as Failure", in Logic and Data Bases. (H. Gallaire and J. Minker Eds.), Plenum Press, New York, 1978, 293-322.

Clark K.L., Tarnlund S.-A., (Eds.), Logic Programming, Academic Press, New York, 1982.

Colmerauer A., "Les System-Q ou un formalisme pour analyser et synthetiser des phrases sur ordinateur", Int. Rep. n. 43, Dept. dInformatique, UniversitÄ di Montreal, 1973.

Constable R.L., Implementing Mathematics with the Nuprl Proof Development System, Prentice-Hall, 1986.

Coquand T., Huet G., "Constructions: a Higher Order Proof System for Mechanizing Mathematics", INRIA Report N. 401, 1985.

Coscia P., Franceschi P., Levi G., Sardu G., Torre L. "Metalevel Definition and Compilation of Inference Engines in the EPSILON Logic Programming Environment", Proc. of Fifth Symposium and Conference on Logic Programming, Seattle, 1988, 359-373.

Coscia P., Franceschi P., Levi G., Sardu G., Torre L. "Object level reflection of inference rules by partial evaluation", in (Maes & Nardi, 1988).

Cousineau G., Curien P.L., Mauny M., "The Categorical Abstract Machine", in Functional Programming Languages and Computer Architecture, J. P. Jouannaud (ED.), Springer-Verlag LNCS 201, 1985, 50-64.

DeGroot D., Lindstrom G., Logic Programming: Functions, Relations and Equations, Prentice-Hall, 1986.

Dershowitz N., Plaisted D.A., "Logic Programming cum Applicative Programming", Proc. 1985 Symp. on Logic Programming, IEEE Comp. Society Press, 1985, 54-66.

des Rivieres J., Levesque H.J., "The Consistency of Syntactical Treatments of Knowledge", Proc. of Theoretical Aspects of Reasoning about Knowledge, 1986, 115-130.

Dincbas M., Van Hentenryck P., Simonis H., Aggoun A., Graf T., Berthier F., "The Constraint Logic Programming Language CHIP", Proc. Int'l Conf. on Fifth Generation Computer Systems, Tokyo, 1988.

van Emden M.H., Kowalski R.A., "The Semantics of Predicate Logic as a Programming Language",JACM 23 (4), 1976, 733-742.

Ershov A., "On the Partial Evaluation Principle", Information Processing Letters, 6.2, 1977, 38-41.

Etherington D., Reiter R., "On Inheritance Hierarchies with Exceptions", Proc. of the National Conference on Artificial Intelligence, Washington, 1983.

Fagin R., Halpern J.Y., "Belief, Awareness, and Limited Reasoning: Preliminary Report", Proc. IJCAI 85, Los Angeles, 1985, 491-501.

Fahlman S., NETL: A System for Representing and Using Real World Knowledge, MIT Press, 1979.

Felty A., Miller D., "Specifying Theorem Provers in a Higher Order Logic Programming Language", Proc. Ninth Conf. on Automated Deduction, Argonne, 1988.

Fikes R.E., Hendrix G., "A Network Based Knowledge Representation and its Natural Deduction System", Proc. IJCAI 77, Cambridge, 1977, 235-246.

Fikes R.E., Kehler T.P., "The Role of Frame-Based Representation in Reasoning", CACM Vol. 28, No. 9, 1985.

Fitting M., "Enumeration Operators and Modular Logic Programming", J. of Logic Programming, 4, 1987, 11-21.

Forgy C.L., "Rete: a fast algorithm for the many pattern / many object pattern match problem", Artificial Intelligence 19 (1), 1982.

Fribourg L., "SLOG: A logic programming language interpreter based on clausal superposition and rewriting", Proc. 1985 Symp. on Logic Programming, IEEE Comp. Society Press, 1985, 172-184.

Furukawa K., Takeuchi A., Kunifuji S., Yasukawa H., Ohki M., Ueda K., "Mandala: A logic based knowledge programming system", Proc. Int'l Conf. on Fifth Generation Computer Systems, 1984, 613-622.

Gallagher J., "Transforming Logic Programs by Specializing Interpreters", Proc. ECAI 86, Brighton, 1986.

Gallaire H., "Merging Objects and Logic Programming: Relational Semantics", Proc. AAAI-86, Philadelphia, 1986, 745-758.

Gallaire H., Minker J., Logic and Data Bases, Plenum Press, New York, 1978.

Gallaire H., Minker J., Nicolas J., "Logic and Databases: A Deductive Approach", Comp. Surveys, 16, 2, (1984), 153-185.

Genesereth M.R., Nilsson N.J., Logical Foundations of Artificial Intelligence, Morgan Kaufmann Publishers, 1987.

Giordano L., Martelli A., Rossi G.F., "Local Definitions with static scope rules in logic programming", Proc. Int'l Conf. on Fifth Generation Computer Systems, 1988, 389-396.

Goguen J.A., "One, None, a Hundred Thousand Specification Languages", Proc. Information Processing 86, Dublin, 1986, 995-1003.

Goguen J.A., Burstall R., "Institutions: Abstract Model Theory for Computer Science", Centre for the Study of Language and Information, Stanford, Rep. N. CSLI-85-30, 1985.

Goguen J.A., Meseguer J., "Equality, Types, and (why not?) Generics for Logic Programming", J. of Logic Programming, 1, 1984, 179-210.

Goguen J.A., Meseguer J., "Unifying Functional, Object-Oriented and Relational Programming with Logical Semantics", Centre for the Study of Language and Information, Stanford, Rep. N. CSLI-87-93, 1987.

Gordon M.J., Milner A.J., Wadsworth C.P., "Edinburgh LCF: A Mechanized Logic of Computation", LNCS Vol.78, Springer-Verlag, 1979.

Gupta A., "Parallelism in Production Systems: the Sources and the Expected Speed-up", Proc. of 5th International Workshop on Expert Systems and Their Applications, Avignon, 1985.

Halpern Y.H., Moses Y., "A Guide to Modal Logics of Knowledge and Belief: Preliminary Draft", Proc. of 9th IJCAI, Los Angeles, 1985.

Hayes P.J., "Some problems and non-problems in representation theory", Proc. of AISB Summer Conf., Univ. of Sussex, 1974, 63-69, reprinted in (Brachman & Levesque, 1985), 3-22.

Hayes P.J., "In Defence of Logic", Proc. of IJCAI-77, Cambridge (MA), 1977, 559-565.

Hayes P.J., "The Logic of Frames", in Frame Conceptions and Text Understanding, (D. Metzing Ed.), Walter De Gruyter and Co., Berlin, 1979, reprinted in (Brachman & Levesque, 1985), 287-295.

Hill R., "LUSH-Resolution and its Completeness", DCL Memo 78, Dept of Artificial Intelligence, University of Edinburgh, 1974.

Hintikka J., Knowledge and Belief, Cornell University Press, 1962.

Huet G.P., "A Unification Algorithm for Typed λ-Calculus", Theoretical Computer Science 1, 1975, 27- 57.

Jaffar J., Lassez J.L., "Constraint Logic Programming", Proc. of POPL 87, Munich 1987.

Jaffar J., Lassez J-L., Lloyd J.W., "Completeness of the Negation as Failure Rule", IJCAI 83, Karlsruhe, 1983, 500-506.

Jaffar J., Lassez J.L., Maher M.J., "A Theory of Complete Logic Programs with Equality", J. of Logic Programming, 3, (1984) 211223.

Jaffar J., Lassez J.L., Maher M.J., "A Logic Programming Language Scheme", in (DeGroot & Lindstrom, 1986).

Jaffar J., Michaylov S., "Methodology and Implementation of a CLP System", Proc. of Fourth Int'l Conference on Logic Programming, 1987, The MIT Press.

Kahn K., Tribble E.D., Miller M.S., Bobrow D.G., "Objects in Concurrent Logic Programming Languages", in OOPSLA'87, N. Meyrowitz (Ed.), 1986.

Kauffmann H., Grumbach A., "Representing and manipulating knowledge within worlds", Proc. First International Conf. on Expert Data Base Systems, (L.Kershberg, Ed.), 1986, 61-73.

Kauffmann H., Grumbach A., "Representing and manipulating knowledge within worlds", Proc. First Int. Conf. on Expert Data Base Systems, L. Kershberg Ed., 1986, 61-73.

Klop J.W., "Term Rewriting Systems: a Tutorial", Bulletin of the EATCS 32, 1987, 143-182.

Konolige K., "On the Relation Between Default Theories and Autoepistemic Logic", IJCAI 87, Milano 1987.

Konolige K., "Hierarchical Autoepistemic Theories for Non-monotonic Reasoning", Proc. of Second Int. Workshop on Non-monotonic Reasoning, Grassau, 1988.

Kowalski R.A., "Predicate Logic as a Programming Language", Proc. IFIP'74, Stockholm, North Holland, 1974, 569-574.

Kowalski R.A., Logic for Problem Solving, North Holland, New York, 1979.

Kowalski R.A., "Algorithm = Logic + Control", C. ACM 7, 1979,424-435.

Kowalski R.A., Kuehner D., "Linear Resolution with Selection Function", Artificial Intelligence 2, 1971, 227-260.

Kowalski R.A., Sergot M., "A Logic-based Calculus of Events", New Generation Computing, 4, 1, 1986, 65-95.

Kunen K., "Negation in Logic Programming", J. Logic Programming 4, 4, 1987, 289-308.

Kunen K., "Signed Data Dependencies in Logic Programs", University of Wisconsin, J. Logic Programming 7(2), 1989, 231-245.

Kursawe P., "How to invent a PROLOG Machine", Proc. of Third International Conference on Logic Programming, London 1986, 134-148.

Lambek J., Scott P.J., Introduction to Higher Order Categorical Logic, Cambridge University Press, 1986.

Lassez J.L., Marriot K., "Explicit and Implicit Representation of Terms Defined by Counter Examples", J. of Automated Reasoning, 1987.

Levesque H., "A Logic of Implicit and Explicit Belief", in Proc. AAAI-84, Austin, TX, 1984, 502-508.

Levesque H.J., "Making Believers out of Computers", Artificial Intelligence 30, 1986, 81-108.

Levesque H.J., Brachman R.J., "A Foundamental Tradeoff in Knowledge Representation", Proc. CSCSI, London Ontario, 1984, 141-152, reprinted in (Brachman & Levesque, 1985).

Levi G., Sardu G., "Partial Evaluation of Metaprograms in a 'Multiple Worlds' Logic Language", New Generation Computing, 6, 1988, 227-247.

Lifschitz V., "Computing Circumscription", Proc. of 9th IJCAI, Los Angeles, 1985, 121-127.

Lifschitz V., "On the Declarative Semantics of Logic Programs with Negation", Proc. of Workshop on Foundations of Deductive Data Bases and Logic Programming, Washington, 1986, 420-432.

Lloyd J.W., Foundations of Logic Programming. Second Edition, Springer-Verlag, Symbolic Computation Series, 1987.

Maes P., Nardi D., Meta-Level Architectures and Reflection, North-Holland, 1988.

Mancarella P., Pedreschi D., Turini F., "Functional Metalevel for Logic Programming", in (Maes & Nardi, 1988), 1988, 329-344.

McAllister D., "Reasoning Utilities Package User's Manual", MIT AI Lab Memo, 667, 1982.

McCabe F., "Logic and objects - Part one: the Language", Research Report N. DOC87/9, Imperial College, 1988.

McCarthy J., "Circumscription - A Form of Non-monotonic Reasoning", Artificial Intelligence 13, 1980, 27-39.

McDermott D., "Tarskian Semantics, or No Notation Without Denotation!", Cognitive Science 2, 1978, 277-282.

McDermott D., "Non-monotonic Logic II: Nonmonotonic Modal Theories", Journal of ACM 29, 1982, 33-57.

McDermott D., Doyle J., "Non-monotonic Logic I", Artificial Intelligence 13, 1980, 41-72.

Meseguer J., Goguen J.A., Smolka G., "Order-sorted Unification", Tech. Rep. N. CSLI-87-86, Univ. of Stanford, 1987.

Miller D., "A Theory of Modules for Logic Programming", Proc. Symp. on Logic Programming, Salt Lake City, 1986, 106-114.

Miller D., Nadathur G., Scedrov A., "Hereditary Harrop formulas and uniform proof systems", Proc. Symp. on Logic in Computer Science, Ithaca, 1987, 98-105.

Minsky M., "A Framework for Representing Knowledge", in Mind Design, J. Hougeland Ed., MIT Press, Cambridge 1981, 95-128, reprinted in (Brachman & Levesque, 1985).

Moore R.C., "The role of Logic in Knowledge Representation and Common Sense Reasoning", Proc. of AAAI-82, Pittsburgh (PA), 1982.

Moore R.C., "Semantical Considerations on Nonmonotonic Logic", Proc. of IJCAI 83, 272-279.

Moore R.C., "A Formal Theory of Knowledge and Action", in Hobbs J., Moore R.C. (eds.) Formal Theories of the Common sense World, Ablex Pub. Corp., 1985, 455-458.

Montague R., "Syntactical Treatment of Modality, with Corollaries on Reflexion Principles, and Finite Axiomatizability", Acta Philos. Fenn., 16, 1963, 153-167.

Mycroft A., OKeefe R.A., "A Polymorphic Type System for PROLOG", Artificial Intelligence, 23, 1984, 295-307.

Nadathur G., Miller D., "An introduction to λ-PROLOG", Proc. of Fifth Symposium and Conference on Logic Programming, Seattle, 1988.

Nakashima H., "Knowledge Representation in PROLOG/KR", Proc. Int. Symposium in Logic Programming, Atlantic City, 1984, 126-130.

Nait Abdallah M.A., "Procedures in Horn Clause Programming", Proc. Third Int. Conf. of Logic Programming, London, 1986, LNCS 255, 433-447.

Nait Abdallah M.A., "Logic Programming with Ions", Proc. ICALP 87, Karlsruhe, 1987, 11-20.

Nait Abdallah M.A., "Heuristic Logic and the Process of Discovery", Proc. of Fifth Symposium and Conference on Logic Programming, Seattle, 1988, 859-875.

Nebel B., "Computational Complexity of Terminological Reasoning in BACK", Artificial Intelligence Journal, 34 (3), April 1988.

Nelson G., Oppen D.C., "Simplification by Cooperating Decision Procedures", ACM Transactions on Programming Languages and Systems, Vol.1, 2, October, 1979.

O'Donnell M., Equational Logic as a Programming Language, M.I.T. Press, 1985.

OKeefe R.A., "Towards an Algebra for Constructing Logic Programs", Proc. Symp. on Logic Programming, Boston, 1985, 152-160.

Patel-Schneider P., "Small can be Beautiful in Knowledge Representation", Proc. of IEEE Workshop on Principles of Knowledge Based Systems, Denver, Colorado, IEEE Computer Society, Dec 84.

Patel-Schneider P., "A Four-Valued Semantics for Terminological Logics", Artificial Intelligence. 38, 1989, 319-351.

Patel-Schneider P., "Undecidability of Subsumption in NIKL", Artificial Intelligence 39, 1989, 263-272.

Paulson L.C., "Natural Deduction as Higher Order Logic", Journal of Logic Programming, 3, 1986, 237-258.

Pereira F., "Logic Programming", Journal of Automated Reasoning 1,1, 1985, 9-13.

Perlis D., "Languages with Self-Reference I: Foundations", Artificial Intelligence, 25, 1985, 301-322.

Perlis D., "Languages with Self-Reference II: Knowledge, Belief and Modality", Artificial Intelligence, 34, 2, 1988, 179-212.

Plotkin G.D., "Building-in Equational Theories", Machine Intelligence 7, Halsted, Wiley, NY, 1972, 73-90.

Przymusinski T.C., "On the Semantics of Stratified Deductive Databases", Proc. Workshop on Foundations of Deductive Data Bases and Logic Programming, Washington, 1986, 433-443.

Przymusinski T.C., "Non-monotonic Reasoning vs. Logic Programming: a New Perspective". in Formal Foundations of AI, D. Partridge and Y. Wilks eds., Cambridge University Press, London, 1991.

Quillian M.R., "Word Concepts: a Theory and Simulation of Some Basic Semantic Capabilities", Behavioural Science 12, 1967, 410-430, reprinted in (Brachman & Levesque, 1985).

Reiter R., "On Closed World Data Bases", in Logic and Data Bases (Gallaire, H. and Minker,J. Eds.) Plenum, New York, 1978, 5576.

Reiter R., "A Logic for Default Reasoning", Artificial Intelligence 13, 1980.

Reiter R., "Circumscription implies Predicate Completion", Proc. of AAAI-82, Pittsburgh, Morgan-Kauffman, 1982, 418-420.

Rich C., "The Layered Architecture of a System for Reasoning about Programs", IJCAI-85, Los Angeles, 1985.

Roberts R.B., Goldstein I.P., "The FRL Manual", AI Memo 409, MIT AI-Lab, 1977.

Robinson J.A., "A Machine-oriented Logic Based on the Resolution Principle", J. ACM 12, 1, 1965, 23-41.

Safra S., Shapiro E., "Meta-interpreters for real", Information Processing-86, H.-J. Kugler, Ed., Elsevier Science Publishers B.V., 1986.

Sandewall E., "Nonmonotonic Inference Rules for Multiple Inheritance with Exceptions", Proc. of IEEE, Vol. 74, no 10, 1986, 1345-1353.

Shapiro E., Takeuchi A., "Objects Oriented Programming in Concurrent Prolog", in New Generation Computing 1, 1983.

Shepherdson J.C., "Negation as Failure: a Comparison of Clark's Completed Data Bases and Reiter's Closed World Assumption", J. of Logic Programming, 1,1, 1984, 5179.

Shepherdson J.C., "Negation as Failure II", J. of Logic Programming, 2,3,1985, 185202.

Shepherdson J.C., "Negation in Logic Programming", in Foundations of Deductive Databases and Logic Programming (J. Minker Ed.), Morgan Kaufman, Los Altos, 1987.

Simi M., Motta E., "Omega: an Integrated Reflective Framework", in (Maes & Nardi, 1988).

Stabler E.P., "Object-Oriented programming in PROLOG", AI Expert, 1986, 46-57.

Steels L., "The KRS Concept System", Vrjie Universiteit Brussel, Artificial Intelligence Lab. Tech . Rep. 86-1, Brussels, 1985.

Sterling L., "Meta-interpreters for expert systems", CAISR TR 134-85, Case Western Reserve University, 1985.

Sterling L., "A Metalevel Architecture for Expert Systems", in (Maes & Nardi, 1988).

Sterling L., Beer R.D., "Incremental flavour-mixing of meta-interpreters for expert system construction", Proc. 1986 Symp. on Logic Programming, IEEE Comp. Society Press, 1986, 20-27.

Sterling L., Shapiro E., The Art of Prolog, MIT Press, 1986.

Stickel M.E., "Automated Deduction by Theory Resolution", Journal of Automated Reasoning 1, 1985, 333-355.

Takeuchi A., Furukawa K., "Partial evaluation of PROLOG programs and its application to metaprogramming", Information Processing-86, (H.-J. Kugler, Ed.) Elsevier Science Publishers B.V., 1986, 415-420.

Touretzky D.S., The Mathematics of Inheritance Systems, Morgan-Kaufmann, Los Altos CA, 1986.

Vilain M., "The Restricted Language Architecture of a Hybrid Knowledge Representation System", Proc. IJCAI-85, Los Angeles, 1985.

Weyhrauch R.W., "Prolegomena to a Theory of Mechanized Formal Reasoning", Artificial Intelligence 13, 1980, reprinted in (Brachman & Levesque, 1985).

Woods W.A., "What's in a Link: Foundations for Semantic Networks", in Representation and Understanding: Studies in Cognitive Science, D.J. Bobrow and A.M. Collins Eds., Academic Press, New York, 1975, 35-82, reprinted in (Brachman & Levesque, 1985).

Zaniolo C., "Object-Oriented programming in PROLOG", Proc. Int. Symposium on Logic Programming, Atlantic City, 1984.

CHAPTER 2

Language Understanding by Computer: Developments on the Theoretical Side

Harry Bunt
ITK, Institute for Language Technology and AI, Tilburg, The Netherlands

1. INTRODUCTION

This paper consists of three parts. In the first part I discuss the notion of language understanding and how it relates to Artificial Intelligence. In the second part I review some of the more important recent work on the theoretical side in the design of computer systems intended to understand natural language. In the third part I present a view on directions in the computational modelling of language understanding that seem most important for the near future.

2. UNDERSTANDING LANGUAGE

2.1 Human and Artificial Language Understanding

Until two decades ago, the only type of language understander was the human understander; *language understanding* was synonymous with human language understanding, and the study of language understanding was part of cognitive psychology and psycholinguistics. In the sixties, Chomsky pointed out the theoretical importance of the fact that humans are able to understand infinite varieties of natural-language expressions in spite of finite information-processing resources; the implication being that meaning is encoded in natural language in systematic ways, describable by finite sets of grammatical rules and principles in combination with lexical knowledge.

Since computers are able to store and effectively apply lexicons and large sets of rules in complex tasks, the human understander is no longer the only conceivable kind of language understander. When undertaking the design of a language understanding system, we have to face the question of what it is exactly that has to happen inside the machine in order to speak of "understanding". In other words, what exactly should be the result of an understanding process. This question does not arise so urgently in the case of human language understanding, since it is usually rather obvious whether someone understands something or not. But the computer case is different, as the classical ELIZA program testifies: even when a computer responds in a seemingly intelligent fashion to natural language inputs, it is far from certain that the system actually understands. To determine this, we should consider the system's potential responses to potential inputs, rather than its actual responses to particular inputs. As a system's potential responses are determined by the internal state that is created by the processing of an input, the internal state is what we should look at. One of the major attractions of the study of computational language understanding is precisely this: we can directly inspect internal states of the system, in contrast with the case of a human language understander.

2.2   Understanding and Meaning representation

In practice, language understanding systems are designed so as to produce symbolic structures supposed to represent meaning, so-called "meaning representations". This raises the question what makes these structures representations of *meaning*; what criteria do we have to determine whether the construction of these structures indeed amounts to *understanding*?

For one thing, these structures should themselves have well-defined meanings, otherwise they can hardly explain much about the meanings of natural-language expressions. For this reason we shall in this paper only consider approaches to computational language understanding that use rigorously defined meaning representations. Another requirement is that these structures should have the logical properties necessary to explain the semantic relations between natural-language expressions. Early work in language understanding by computer has been unsatisfactory in these respects. For instance, the structures produced by the interpretation module in Winograd's pioneering system SHRDLU (Winograd, 1972) consist of pieces of LISP code which do not permit general reasoning. The system is therefore unable to answer or even interpret a question like "What is the colour of a red block" if it cannot find a red block. Schank's "conceptual dependency" structures have no formal definitions at all, let alone the required logical properties.

The use of explicit meaning representations may be compared to the role of semantic representations in linguistic semantic theories. For a long time, linguistic theories have remained vague about the nature of semantic representations. The development of Montague grammar in the early seventies changed this dramatically, however. Montague's work has popularized the use of expressions in a formal, logical language as semantic representations. In Montague's view, the use of these representations is merely a matter of convenience: in principle, the rules for associating them with natural-language expressions could be combined with the rules for their evaluation, giving rise to more complex rules that assign semantic values directly to natural-language expressions. In this approach, meanings ("semantic values") are abstract, non-symbolic objects, modelled by mathematical constructs like functions from possible worlds and timepoints to sets of individuals. This is the most popular view on meanings and meaning representations in linguistic semantic theories. A notable exception is Kamp's Discourse Representation Theory (Kamp, 1981), which regards both semantic representations and abstract model-theoretic constructs as essential ingredients of a theory of linguistic meaning.

2.3   Meaning Representations and Meanings

Can a computer understand language without computing meanings, by only computing meaning representations? The answer is no, as the following example shows.

Suppose a computer is asked the question: "Which flights from Canada arrive on Monday?" Let's assume that a meaning representation indicates that we are dealing with a question, and that its "content" is the set of flights from Canada arriving on Monday: (Note 1)

(1) <QUESTION,
      {x ∈ FLIGHTS | FROM(x, canada) & ARRIVE(x, monday)}>

In order to answer this question, the machine has to identify those objects which are flights, which depart from somewhere in Canada, and which arrive on Monday. Doing so is applying the semantic definition of the representation language and evaluating the expression which represents the semantic, model-theoretic content. This is in fact the same as computing the meaning of the question. To answer a question, we might say, the machine has to compute its meaning. This is not as obvious as it may seem, however. What is obvious is that a question has to be *understood* before it can be answered, but understanding could conceivably consist of constructing a meaning representation, rather than computing the meaning. The example makes clear, however, that a computer can only be said to understand a question if, in addition to constructing a meaning representation, there is also the *ability to evaluate* that representation: to compute the meaning.

Incidentally, we see here the one and only appropriate criterion for deciding the adequacy of an alleged semantic representation: when applying the semantic definition of the representation language, it should yield (see Note 2) the semantic object that actually constitutes the meaning of the original natural-language expression. (Such a "semantic object" may be a real-world object, a number, a truth value, a set of any of these, a relation, an intension, etc.) The only reliable test for whether a computer has understood something is thus to see what happens when it computes the value of the meaning representation that it has constructed.

How can the value of a meaning representation like (1) be computed, when that value is built up from abstract entities (after all, a flight is an abstract concept)? One way is by consulting a data base which contains descriptions of these objects. A data base is commonly viewed as a representation of the abstract or concrete concepts that make up a certain universe of discourse. Instead of dealing with abstract or real-world objects the machine deals with descriptions of them, and so computer understanding of language after all does not leave the level of symbolic representation and computation.

Another possibility is to view the computing system, including its data base, as an information processing system which is in a certain state, depending on the contents of all its memory registers. Understanding a natural-language expression would change the state of the system; for instance, understanding the above question should bring the system into a state where it knows that the speaker wants to know the value of the corresponding semantic representation. On this view, the meaning representation (1) is a description of the fact that the machine's state changes in that particular way (Note 3).

2.4   Sentence Understanding and Knowledge Structures

According to standard linguistic semantic theory, the meaning of a sentence is determined by the meanings of its words plus its syntactic structure. The meaning of a word, moreover, is given by a relation to a real-world or abstract entity. (This applies to *content* words, at least.) The entities involved in the meanings of content words make up the objects in terms of which our knowledge of the world is expressed. The understanding of a word thus amounts to relating it to entities in the interpreter's total knowledge of the world, and understanding a sentence can be characterized as assigning it a semantic structure in terms of the world knowledge accessed by word meanings. This means that the study of language understanding by computers involves both linguistic semantics, as the study of the linguistic encoding of semantic structures, and Artificial Intelligence, as the study of the representation and application of knowledge.

2.5 Language Understanding and Understanding Language Use

When *studying* language, one may be dealing with *sentences*: word sequences with certain grammatical properties; when *using* language, by contrast, one deals with *utterances*: sentences produced by someone and addressed to someone in a certain context, with a certain intonation or punctuation, and meant to serve a certain purpose. This means that understanding language "in action" is not "just" a matter of sentence semantics, but also involves understanding the purpose of its use in a given context. The full meaning of an utterance has, besides a semantic part, a pragmatic part which describes its communicative functions (cf. (1) above).

"Context" should be taken in a double sense here: linguistic and nonlinguistic. The linguistic context is the discourse of which the utterance forms part. "Nonlinguistic context" refers to a variety of factors, including those concerning the *setting* in which language is used and those concerning the *states* of the agents involved, which include their information, plans, goals, hopes, fears, etc. Both linguistic and nonlinguistic context play an essential part in establishing the meaning of an utterance.

Let us look at an example of actual language use to see this. Text No. 2 is a transcription of a telephone conversation with the information service at Schiphol, Amsterdam Airport (C = client, I = information service).

(2)

1. I: Schiphol information.
2. C: Good afternoon.
   This is Van I. in Eindhoven. I would like to have some information about flights to Munich. When can I fly there between now and ... next Sunday.

   3. *I*: Let me have a look. Just a moment.
   4. *C*: Yes.
   5. *I*: O.K., there are ... three flights every day, one at nine fifty.
   6. *C*: Yes.
   7. *I*: One at one forty ... and one at six twenty-five.
   8. *C*: Six twenty-five ... These all go to Munich.
   9. *I*: These all go to Munich.
  10. *C*: And that's on Saturday too.
  11. *I*: That's on Saturday too, yes.
  12. *C*: Right ... Do you also have information about the connections to Schiphol by train?
  13. *I*: Yes, I do.
  14. *C*: Do you know how long the train ride takes to Schiphol?
  15. *I*: You are travelling from Eindhoven?
  16. *C*: That's right.
  17. *I*: It's nearly two hours to Amsterdam ... You change there and then it's another fifteen minutes, so you should count on some two and a half hours.
  18. *C*: O.K. ... Thank you.
  19. *I*: You're welcome.
  20. *C*: Good afternoon.
  21. *I*: Good afternoon.

The first three sentences: "Schiphol information", "Good afternoon", "This is Van I. in Eindhoven" can clearly not be understood without considering their role and place in the dialogue (and its setting). In fact, their role and place are their meaning in this case: their primary function is to assess the contact established, checking the most elementary conditions for starting a dialogue. (Compare the "Good afternoon" in turn 2 with those in turns 20 and 21!). Likewise, the "Yeses" in 4, 6, 11, and 13 can of course be interpreted only in relation to the previous discourse.

Other cases where contextual information is needed to establish the communicative function are the sentences "These all go to Munich" and "And that's on Saturday too", in turns 8 and 10. The transcriptions of these sentences do not contain question marks, since the utterances did not have an obvious "interrogative" intonation. A context-independent analysis of these sentences cannot possibly reveal that they function as questions. Note, in particular, that the same sentence occurs in turn 8 as a verification and in turn 9 as a confirmation. The question as to what is the function of an utterance in the dialogue arises in fact for *every* utterance; therefore, the semantic analysis of the sentences should all the time be supplemented with a pragmatic one which relates the sentence to the current context.

The example also illustrates the necessity to extend interpretation over sentence boundaries in order to establish their *semantic* content. Text No. 3 lists those sentences in the dialogue that deserve a serious semantic analysis.

(3)

1. I would like to have some information about flights to Munich.
2. When can I fly there between now and ... next Sunday?
3. There are ... three flights every day, one at nine fifty one at one forty ... and one at six twenty-five.
4. These all go to Munich.
5. That's on Saturday too.
6. Do you also have information about the connections to Schiphol by train?
7. Do you know how long the train ride takes to Schiphol?
8. You are travelling from Eindhoven?
9. It's nearly two hours to Amsterdam.
10. You change there and then it's another fifteen minutes, so you should count on some two and a half hours.

Sentence 1 can be interpreted in isolation, but cannot get the *intended* interpretation, where the quantification domain is restricted to flights from Amsterdam. Sentence 2 suffers from the same problem. Sentence 3 even more so; interpreted in isolation, it represents an obviously *false* assertion, where it is true in its intended interpretation! Similar cases of incompleteness, or *ellipsis*, occur in 7, 9, and 10.

Linguistic context is, by definition, required for the interpretation of anaphoric expressions, expressions that refer to something mentioned elsewhere in the discourse. Clear cases in this text are "that" in 4 and 5, and "there" in 10. Less clear cases are "too" in 5 and "also" in 6. Other context-dependent expressions are, for instance, "I" in (1), "now" in (2), and "you" in (6), (7) and (10).

Ellipsis and anaphora occur all over the place in dialogues of the kind that one might like to have with a computer, and are only two of the wide variety of phenomena that require information from both the linguistic and nonlinguistic context.

3. LANGUAGE UNDERSTANDING AND AI SYSTEMS

From the above analysis we can conclude that language understanding in machines is a process where formal representations of meaning are constructed and evaluated against a body of linguistic and nonlinguistic knowledge. The main theoretical issues in designing language understanding systems therefore relate to:

1. The definition of adequate and computationally tractable meaning representations, taking both semantic and pragmatic meaning aspects into account;
2. The use of pragmatic and discourse-contextual knowledge in constructing meaning representations; and
3. The representation and use of nonlinguistic knowledge in constructing and evaluating meaning representations.

Focusing on these three issues, I consider in this section some contributions to the study of language understanding from recent work in Artificial Intelligence, in particular in Europe.

3.1   Semantic Representation

The design of semantic representation formalisms is difficult because, on the one hand, natural language allows the expression of a very rich variety of semantic structures, whose representation calls for a highly expressive representation language. On the other hand, the more expressive the representation language, the greater the danger that its logical properties become so complex that the evaluation of its expressions becomes computationally intractable (Note 4.)

For instance, it is well known that many semantic phenomena in natural language can only be handled in an intensional framework. Such frameworks have become well-established in linguistic semantics through the work of Montague, who designed an intensional semantic representation language with a formal semantics, called IL (Intensional Logic), which permitted a successful attack on a variety of semantic problems. The computation of the values of IL expressions, however, presupposes the explicit availability of specifications of all possible combinations of facts in the domain of discourse; though mathematically elegant, for a semantic domain of realistic size this is computationally intractable.

For similar reasons, most of the AI-work on language understanding uses semantic representations in only limited extensions of *first-order* logic, thus prohibiting things like predicates of sets of individuals, functions from predicates to predicates, etc. This is equally unsatisfactory as the lack of intensions, for instance for treating adverbs. In most semantic representation systems the computation of the value of the representation is a combination of consulting a world model and applying postulates that express dependencies between predicates. The application of these postulates requires an implemented deduction system, and computationally tractable deduction systems are not available for unrestricted higher-order logics.

One of the extensions to first-order logic that is gaining wide acceptance consists in making the logic *many-sorted*. This means that the individual objects inhabiting the semantic domain are subcategorized into a variety of "sorts", and the expressions referring to individuals are *typed* in order to indicate what sort of individual is denoted. Argument positions of predicates can then be labelled with sorts, to be able to check that predicates are applied to appropriate types of arguments. This has been implemented in several of the more ambitious projects in Europe that aim at the construction of language understanding systems. One case in point is the LILOG (Linguistics and Logic) project in Germany, carried out by IBM Heidelberg and a group of universities (see Herzog et al., 1986). In this project, a semantic representation language has been defined called L_{LILOG}, which is basically that of many-sorted first-order predicate logic, with a few notable extensions (see Beierle et al., 1988).

In L_{LILOG}, a more complex notion of sorts than the standard one is defined, in that (1) the collection of sorts is partially ordered, and (2) sorts can be structured objects, rather than atomic (just as syntactic categories are nowadays usually viewed as feature bundles, rather than just names; this has in fact inspired the L_{LILOG} design). Two ways of defining structured sorts are illustrated by the following examples:

(4) red-building == *building(colour* : {*red*})
      person == (*father* : *person(name* : *C* : *string*),
           name : *C* : *string*)

The first defines the sort *red-building* as the sort *building* restricted by the feature *colour* : {*red*}. The denotation of an expression of the sort *red-building* would thus be a building with a red colour. The second defines the sort *person* recursively as the property of having a father and a name, where the father is again of the sort person and the name of the father is identical to that of the son.

In L_{LILOG} sorts are treated as part of the language; thus, a sort declaration like:

(5) John ∈ *student*

where *student* is a sort name, is an *LLILOG* expression saying that John is a student.

Sorts also play a role in the TENDUM project in the Netherlands (see Bunt et al., 1985), where a family of semantic representation languages has been designed called the EL family, for *Ensemble-theoretical Language* family. The semantics of these languages is based on ensemble theory, an extension of Zermelo-Fraenkel set theory with objects that have a part-whole structure much like sets, but that have no elements. This extension was designed in order to deal with the semantics of mass terms (Bunt, 1985). For the rest, the EL languages are fully typed, many-sorted and higher-order; their design is based on the semantic representation languages developed earlier in the PHLIQA question-answering project at Philips Research (see Bronnenberg et al., 1980; Scha, 1983). A sophisticated system of complex types is used in these languages, which are expressed in separate members of the EL family: the EL type languages.

The EL languages are also used in the SPICOS project, a joint project of Siemens and Philips aiming at the development of a documentary data base consultation system with natural language input and output in spoken form (see van Deemter et al., 1985; Thurmair, 1987).

In NATTIE, a natural language project in Britain carried out by SRI Cambridge and Cambridge University, a system is built for translating English sentences into formal representations of their literal meanings; this system is called the Core Language Engine (CLE). In the CLE representation language, many-sorted first-order logic is again the point of departure; lambda abstraction and several other features have been added to this.

The CLE system uses two intermediate levels of linguistic analysis between the natural language sentence and its meaning representation (see Alshawi et al., 1987; Alshawi, 1990). These are a syntactic analysis and a "quasi-logical form", which represents the literal sentence meaning except that the scopes of quantifiers and the references of anaphoric expressions are unspecified. For instance, the sentence *A bishop visited every college* is represented at logical form level by the expression:

(6) *quant(exists,A,[event,A],*
    [visit1,A,qterm(a1,B,[bishop1,B]),
    qterm(every1,C,[college1,C])])

The *qterm* expressions here represent unscoped quantifiers. A quantifier scoping algorithm, applied in the stage of constructing the full-fledged meaning representations, finds two possible scope assignments in this example:

(7) *quant(exists,B,[bishop1,B],*
     quant(forall,C,[college1,C],
       quant(exists,A,[event,A],[visit1,A,B,C])))

   quant(forall,C,[college1,C],
     quant(exists,B,[bishop1,B],
       quant(exists,A,[event,A],[visit1,A,B,C])))

It is tempting to consider the "quasi-logical forms", like (6), as kinds of meaning representations which are underspecified in certain respects, namely quantifier scopes. However, the language in which quasi-logical forms are represented in CLE only has a syntactic definition and its semantics is by no means obvious; therefore, these do as yet not qualify as proper meaning representations.

There are good reasons for being interested in formal meaning representations which are underspecified in certain respects, since the use of formal representation languages sometimes has the undesirable effect that one is forced to be more articulate than natural language warrants. Consider, for instance, the sentence: "The boys carried the boxes upstairs". The sentence leaves open whether individual boys carried individual boxes, or piles of boxes, or groups of boys carried individual boxes, etc. Since the sentence asserts something about *the boys* and *the boxes*, rather than *some (of the) boys/boxes*, a semantic representation should somehow quantify universally over the sets of boys and boxes under consideration. But should these quantifiers relate to individual boys and boxes, or to sets of them? A predicate-logic representation language forces us to make specific choices in these matters, and thus to treat the sentence as multiply ambiguous, although intuitively it isn't ambiguous at all; rather, it is *vague* in some respects. Only if this vagueness is taken over in the formal representation, one may truly claim to represent what the sentence expresses. (See further Bunt, 1984.) The recently started ESPRIT project PLUS ('Pragmatics-based Language Understanding System') takes this point as a central element in its design philosophy, aiming at a 'natural language engine' which assigns underspecified semantic representations to inputs, whose interpretation is strongly context dependent (see Black et al., 1991).

An interesting aspect of the CLE representation language, illustrated by (5), is that state/event variables are used as arguments of predicates.

A verb is treated not as expressing a relation between its subject and its complements, but, following Davidson (1967), as describing an event (or state) where that relation holds. This opens the possibility of treating optional verb phrase modifiers as predications of events, which in turn permits a uniform treatment of prepositional phrases, independent of whether they modify nominal or verb phrases. This approach is also followed in the representation language CML (Conceptual Modelling Language) of the LOQUI system, a natural language interface to databases developed as part of the ESPRIT project LOKI (see Imlah, 1987), and in the representation language used in the ACORD project, a mainly Scottish-German ESPRIT project (see Calder, Klein, Moens and Zeevat, 1987).

However, some verb phrase modifiers cannot be handled adequately in this way. For instance, fast swimming goes much slower than fast running; therefore, one would like to interpret the adverb *fast* relative to the action involved. For this purpose, Pulman (1987) has proposed the addition of higher-order predicates applicable to state/event variables. For example, the sentence *Mary is swimming fast* would be represented as:

$$(8) \quad \exists e : SWIM(mary, e) \ \& \ FAST(e, \lambda e' : \exists x : SWIM(x, e'))$$

which can be read as: "There is a swimming by Mary and, by the standards of those events where someone swims, that event was fast".

This extension has not been implemented so far. Pulman suggests that:

> "*in an actual implementation, we can recast the effects of such a function* (like *fast*, HB) *in terms of meaning postulates, or rules of inference, at least in the cases where the property in question is capable of being dealt with in a quantitative way. So we might encode something like "tall" as:*

$$(9) \quad GREATER\text{-}THAN(height\text{-}of(x), 100ft) \rightarrow TALL(x, tree)"$$

This could indeed be done for adjectives like *tall*, but it is unclear how this could work for adverbs like *fast* where it seems wrong to transpose the speed of the event to the speed of its subject. (For example, when *Mary moves her hand fast* it is the object of the verb rather than the subject to which a qualification of speed applies.)

Most of the representation languages mentioned so far take first-order logic as point of departure. This is by far the most common approach in AI circles, but things are changing. The representation languages in the TENDUM project are based on higher-order type logic, and make use in their interpretation of ideas akin to situation semantics (see Barwise & Perry, 1983; Fenstad et al., 1987). In the ACORD project the representation language is based on discourse representation theory; situation semantics is the basis of work carried out at the University of Oslo (see Vestre, 1987; Colban & Fenstad, 1987). These approaches offer better perspectives for integrating sentence semantics with discourse semantics (see Guenthner, 1988; Fenstad, 1988).

3.2 Pragmatics and Discourse

The importance of both linguistic and nonlinguistic context for understanding an utterance in natural language was stressed in 2.5. In fact, contexts and utterances are related in a similar way as hens and eggs, since contexts give rise to utterances and utterances create contexts. Utterances create *linguistic* contexts by definition; less obvious is the fact that utterances also create *nonlinguistic* contexts.

Linguistic communication occurs because some agent wants to achieve certain goals. These goals may vary from specific, well-defined ones such as wanting to know what time it is, to very general ones like creating a pleasant atmosphere. An agent's goals and beliefs form the primary driving forces behind his utterances, and they form part of the nonlinguistic context. It is also this part of nonlinguistic context which is largely *created by* utterances, since the utterances change the agent's beliefs, goals and other considerations; in other words, the utterances *change the context*. The new context which an utterance creates will determine the continuation. Utterances thus produce new contexts, and contexts produce new utterances.

The study of language in relation to context, in particular to nonlinguistic context, takes place in the branch of linguistics called *pragmatics*; the study of language in relation to linguistic context, in particular as far as interpretation is concerned, is called *discourse semantics*. Discourse semantics has in recent years been an area of highly active research by linguists, philosophers, cognitive scientists as well as AI researchers; for an up-to-date review see the chapter by Guenthner in volume 2 of the present series. The AI work in pragmatics which is most relevant to consider here is that where functional aspects of language use are studied in relation to formal and computational representation of mental states and dialogue organization.

When studying the functions of natural-language utterances, it is inviting to adopt the concepts of speech act theory. Speech act theory views the use of language as the performance of acts of communication. Central in this approach is the notion of *illocutionary acts*, being the actions performed in using language. For instance, when we say that a declarative sentence is used to make a request, the request is the illocutionary act performed. Other examples of illocutionary acts are promises, threats, verifications, lies and warnings.

Speech act theory also distinguishes a *perlocutionary* dimension of language use, which concerns the effects that a linguistic act altogether achieves. For instance, convincing someone is not an illocutionary act, but may be achieved indirectly as the effect of one.

Speech act theory has traditionally focused on the development of taxonomies of illocutionary acts and the conditions for their correct performance. This does not immediately bear fruit for the design of language understanding systems. More fruitful is the approach taken by Allwood (1976), where utterances are viewed as actions that signal certain aspects of the speaker's mental state. Such an approach also underlies versions of speech act theory developed in AI by Perrault, Cohen, Allen, Levesque, Bunt, Ostler, and others. In some of these versions, illocutionary acts play an explicit role, in that utterances are assigned illocutionary act labels as part of the representation of their meaning (as in (1) above). In others their role is only instrumental in designing a process that explicates which aspects of a speaker's mental state are signalled by which kinds of utterances (see e.g. Cohen & Levesque, 1985). The most sophisticated versions which can be found in the literature have so far not been implemented; less sophisticated versions are used in the LOQUI system (Wachtel, 1987) and in the TENDUM system (Bunt, 1986).

AI work on speech acts has resulted in significantly deeper insights into the nature of speech act concepts. In a critical review of the foundations of speech act theory, Levinson (1983) suggests that the most promising way of obtaining solid foundations for notions like illocutionary acts consists in viewing these as functions that map contexts into contexts, as suggested earlier e.g. by Gazdar (1981). The crucial question is then, of course, how a notion of context can be defined to make this idea operational. The AI work on speech acts gives an answer to this question: context can be construed as the relevant aspects of the mental states of the participants in the communication. Which aspects of the mental states are "relevant" is determined by the overall setting in which the communication takes place. For example, in pure information-exchange dialogues the relevant aspects include knowledge and belief with various gradations of certainty, as well as goals and plans, but exclude emotions like hopes and regrets.

3.3 Language interpretation and knowledge processing

We have seen above that language understanding by its very nature is a combination of determining linguistic structure and applying world knowledge, since the understanding of a sentence involves the identification of the knowledge elements that the linguistic expression refers to. One of the characteristics of the study of language understanding by computer is that the connection between linguistic structure and nonlinguistic knowledge is made explicit and brought within the scope of investigation.

Apart from the role that nonlinguistic knowledge plays in providing the roots for the analysis of meaning, nonlinguistic knowledge is also needed to make linguistic interpretation feasible. One of the most striking properties of natural language expressions is their ambiguity, both at word level and at sentence level. By far the majority of words have a variety of possible meanings, and by far the majority of sentences a variety of possible interpretations due not only to lexical ambiguities, but also to different scope assignments to quantifiers, conjunctions, temporal adjuncts, etc.; due to different choices of grammatical function, of attaching prepositional phrases, and so on. For human understanders, most of this ambiguity goes virtually unnoticed because the application of world knowledge at an early stage makes one reading more plausible than the others. The design of artificial understanding systems makes us acutely aware of both the perplexing degree of ambiguity in natural language and of the absolute necessity of applying world knowledge at an early stage.

From a logical point of view, there are two methods of combining knowledge elements: the *deductive* and the *model-theoretic* method, also known as the *syntactic* and the *semantic* method, respectively. The deductive method considers the facts about the domain of discourse, assumed to be true, as axioms. Rules of inference are applied to these axioms plus the axioms of logic to try to deduce the truth of certain propositions. The model-theoretic approach works by means of recursive evaluation of complex formulae combined with the assignment of semantic values to the constants and variables of the language in which the formulae are expressed.

We have seen above that most language understanding systems in AI, to the extent that they use formal meaning representations, stick to first-order predicate logic or a modest extension of it. In that case, deductive methods can be applied; for more complex representation languages the deductive method runs into serious difficulties because of its computational complexity. As semantic representation languages become more and more powerful, the model-theoretic approach seems to be the only one possible. This approach has other limitations, however (see 4.3).

The interplay of linguistic and nonlinguistic knowledge would obviously be facilitated by expressing both in the same representation formalism. This would open the way to applying deductive or model-theoretic computations to combinations of linguistic and nonlinguistic elements. A step in this direction is taken in the L_{LILOG} language, which has inherited for the definition of complex sorts as in (4) constructions from a formalism designed for the representation of syntactic feature structures, the STUF formalism (see Smolka, 1988; Bouma, König & Uszkoreit, 1988). Similarly, in the TENDUM project the design is under way of a language in the EL family for describing syntactic structures, feature operations, and syntactic-semantic grammar rules.

4. FUTURE RESEARCH

In the previous section we have focused on three issues in the computer understanding of natural language:

1. the definition of adequate and computationally tractable meaning representations;

2. the use of pragmatic and discourse-contextual knowledge in constructing meaning representations;

3. the representation and use of nonlinguistic knowledge in constructing and evaluating meaning representations.

The development of artificial language understanding systems requires further research on each of these topics; such research may, on the other hand, be expected to contribute significantly to our understanding of language understanding. The design of language processing systems moreover by necessity involves the *integration* of a wide range of theories concerning the many aspects of language. A decently designed language understanding system should not be based on separate theories of semantics, pragmatics, discourse, knowledge representation and reasoning, but on a combination of such theories, brought together in a unifying framework. Below I briefly consider the future of research on each of the three topics mentioned above as well as the possibility of a unifying framework.

4.1   Semantic representation

Within the wider perspective of language understanding in context, taking discourse context and nonlinguistic knowledge into account, classical semantic sentence analysis may almost seem a futile business. Indeed, it is fair to say that a considerable number of studies of semantic problems at sentence level have been misdirected, since the key to their solution is to look beyond the sentence boundary, at discourse level, and to bring nonlinguistic contextual knowledge to bear. Nonetheless, the adequate representation of the semantic content encoded by syntactic structures remains a task of obvious fundamental importance. And there is still a lot of work to be done in this area, especially in refining representation formalisms in a way that supports the systematic construction of semantic representations for natural-language sentences.

The following example illustrates the kind of problems to be addressed.

Suppose we want to construct a semantic representation of the sentence "The people who presented the reports on morphology are with Siemens". The use of the definite articles indicates that there is a predication ranging over known domains, and that all the people and reports in these domains are involved. All the people involved are said to be with Siemens, so a universal quantification over the domain of the relevant people is in order. However, the quantification over reports on morphology should not be universal, since the relative clause is not meant to select those people who presented every report on morphology, but rather those who presented *some* of those reports. The problem is that, like in the notorious "donkey sentences", the same determiner seems to require a universal quantifier in some syntactic positions and an existential one in others. But the problem is more serious: if we use an existential quantification here, and construct a representation like (10), we miss the point that the sentence states something about all the reports on morphology.

$$(10) \quad \supset x \in \{y \in PEOPLE \mid \exists z \in MORPHREPORTS : PRESENT (y,z)\}: WITH (x, Siemens)$$

The representation (10) would express a true proposition if some of the reports were not presented at all. So in order to represent that all the reports in question are involved, we have to do something to the effect of adding the clause (11):

$$(11) \quad \supset x \in MORPHREPORTS : \exists y \in PEOPLE : PRESENT(y,x)$$

The difficulty here is not so much to give a correct semantic representation of the sentence in first-order logic, but to arrive at such a representation through a systematic procedure based on the syntactic and lexical information provided by the sentence.

It is all too well known that classical first-order logic is not powerful enough for the semantic representation of natural language; the representation of adverbials considered above illustrates this (see also Allen, 1988). The introduction of higher-order predicates or intensions leads to representation languages which are computationally intractable, however. Since the correct representation of semantic structures is impossible in very simple languages, there is really no alternative but to introduce more powerful languages and be very selective in their use in inferential processing. That is, one should not want to use the representations in these languages in general deductive processes of the kind of mechanical theorem provers. The use of appropriate semantic representation formalisms is thus intimately connected with issues of knowledge representation and application, taken up below.

From a linguistic semantic point of view, the hunt should go on for formal representation systems which are powerful enough to represent the semantic distinctions that can be expressed in natural language, without thrusting distinctions upon us which may be relevant from a logical point of view but are not justified on linguistic grounds. Moreover, a useful representation system should allow systematic derivation of representations from linguistic structures.

4.2   Pragmatics and Discourse Semantics

Compared to semantic interpretation at sentence level, pragmatic interpretation at utterance level and semantic interpretation at discourse level are still poorly understood.

Semantic interpretation at discourse level concerns the way in which linguistic context contributes to the determination of sentence semantics. In (3) above we have seen the crucial and pervasive nature of this contribution illustrated in a sample of natural discourse. Anaphora, ellipsis, and the contextual determination of the domain of reference of noun phrases are some of the notorious manifestations of this phenomenon. Ambiguity is another one. For instance, the classical ambiguity concerning the attachment of the preposition phrase in the sentence *He saw a girl with the telescope* vanishes if the sentence is preceded by: *John looked down the dusky street, wondering whether there was somebody at the far end. It was impossible to see. He took the telescope he'd received for his birthday, placed it on the window-sill and looked again.*

Pragmatic interpretation at utterance level concerns the determination of the communicative functions (or "illocutionary forces") of utterances. The AI work on speech act theory, mentioned in 3.2, has so far paid little attention to the question *how* the communicative functions of utterances can actually be determined. It seems rather obvious that, just like the semantic content, the communicative function of an utterance is partly determined by properties of the utterance itself, partly by the linguistic context, and partly by the nonlinguistic context. Exactly which properties of utterances are involved (syntactic structure, lexical choices, special particles, prosodic features, punctuation,..); what contextual features are involved, and how all these factors interact, is still only poorly understood. Some interesting studies in this area, which contribute to the empirical foundations of the kind of speech act theory emerging from AI, are Beun (1988; 1989) and Saebo (1988).

Another aspect of the pragmatic part of language understanding which is still underdeveloped, but beginning to receive attention in AI, concerns the perlocutionary dimension of speech acts. This dimension presents very complex problems, since the perlocutionary effects which an utterance may bring about are hard to predict. They are not "controlled" by the utterance, but depend on properties of the mental states that they interact with. But this does not mean that nothing systematic can be said about perlocutionary effects. For instance, in the setting of an information-exchange dialogue like (2), where one of the participants is considered to be the expert on the topic of conversation, the information on the topic supplied by the expert will in general be taken over by the other participant unless it would conflict with some other information available to that participant. This means that the adoption of beliefs should take place via *defeasible* rules.

Recently, the first steps in developing such rules within a framework of default reasoning have been made by Perrault (1988) and Appelt & Konolige (1988). The formulation of *perlocutionary* rules and the design of appropriate reasoning mechanisms for predicting perlocutionary effects constitute a task of fundamental importance for the development of models of language understanding and language understanding systems, since speech acts are typically performed with the intention to produce a particular perlocutionary effect.

4.3   Linguistic Interpretation and Knowledge Processing

We have seen above that the explicit formal representation of linguistic semantic structures requires a powerful representation language, which poses a problem for effective application of world knowledge - at least when the deductive method is followed. The obvious consequence would seem to be that deductive methods should be avoided; the model-theoretic approach is, by its top-down recursive nature, more directed than the deductive method. However, the model-theoretic method cannot achieve everything through recursive evaluation.

The reason for this is that the model-theoretic approach assumes the value of a complex expression to depend only on the values of its atomic constituents, given the syntactic structure of the expression. But this is not always the case. Suppose we want to express the knowledge that every airplane has an engine, even though we are unable to specify the engine(s) of any particular plane. In this case we know that the formula (12) is true, even though recursive evaluation would fail:

$$(12) \quad \supset x : PLANE(x) \rightarrow \exists y : ENGINE(y) \ \& \ HAS(x,y)$$

Recursive evaluation would say that this formula is *true* in case the expression to the right of the arrow would evaluate to true for every value of x which belongs to the set of planes. That expression would in turn evaluate to true if, for any value a of x, there is a value b of y such that b is an engine which a has. However, we assumed that we didn't know any engine b of any air plane a, so this expression does not come out true. The point is that our knowledge of the truth of the proposition (12) is *independent* of our knowledge of its constituents.

In model-theoretic semantics, this is solved by adding postulates to the knowledge base, which are considered as constraints that the semantic values of the constants and variables of the language should satisfy. But bringing the knowledge expressed by these postulates to bear amounts to performing deductions, and brings us back to the deductive method. All in all, hybrid systems which combine syntactic and semantic methods of inferencing will probably have to be developed.

It was noticed above that the interplay of linguistic and nonlinguistic knowledge would be facilitated by expressing both in the same representation formalism. But it seems unlikely that this is always possible. "Visual" knowledge, such as what a rose looks like, or other sensory knowledge such as how a banana tastes or how honeysuckle smells, does not seem to be representable in symbolic form. Is this kind of knowledge relevant to language understanding?

Until very recently, there was little reason to worry about such matters when building a language understanding system. However, a recent trend discussed in Wahlster's chapter in vol. 2 of the present series, is to make such systems *multimodal*. One may, for instance, ask questions about information acquired through a visual channel, which means that semantic representations will contain symbols that refer to picture elements. Research on such systems may be expected to bring some insights into this largely unexplored domain.

4.4   A Unifying Framework

One of the most important reasons why the design of language understanding systems has a special contribution to make to our understanding of language understanding, is that in such a system many pieces must fit together: syntax, morphology, semantics, pragmatics, discourse semantics, and nonlinguistic knowledge. Moreover, they have to fit together in a computationally feasible way. Within linguistic theory, frameworks have to some extent been constructed where syntax, morphology and semantics fit together; the integration of pragmatics, discourse semantics and nonlinguistic knowledge into these frameworks has not yet been achieved in satisfactory ways. An important task on the theoretical side of designing language understanding systems is the development of a unifying framework in which these pieces fit together at a theoretical level.

Such a framework can be constructed by combining the ideas underlying the AI-approach to speech act theory discussed in 3.2 (including the empirical and perlocutionary extensions mentioned in 4.2) with those underlying current developments in discourse semantics and "dynamic" sentence semantics.

The latter approach, put forward recently by Groenendijk and Stokhof (1987; 1988), applies the interpretation method of Dynamic Logic, developed in the theory of programming languages, to the semantics of classical logical languages like those of predicate logic and intensional logic. When these languages are used for semantic representation, the method is indirectly applied to natural-language semantics.

The idea of Dynamic Logic (see Harel, 1984) is that the meaning of a statement in a programming language is the way the state of a machine changes when the program is run. A computer program is treated as a formal-language expression which contains variables, whose values characterize the state of the machine. Running the program changes the state of the machine as the values of variables change. Thus, if g is the function describing the values of the variables before the program P is run, and h the function describing this afterwards, the pair $<g,h>$ represents a state transition that forms part of the meaning of P. The full meaning of P is the set of all state transitions which, depending on the initial state, may be brought about by P.

This method has been applied to the semantics of first-order logic (Groenendijk & Stokhof, 1987) and Montague's intensional logic

(Groenendijk & Stokhof, 1988). Instead of considering the meaning of a formula φ to be its truth value, relative to a model M and an assignment of values to variables g, the meaning of φ is taken to be the set of all assignment-changes $<g,h>$ such that φ is true relative to M and h. The following example illustrates this:

(13) $| | \exists x : \psi | | = \{<g,h> | h[x]g$ and $h(x) \in | | \psi | | \}$

where the notation '$h[x]g$' means that h is equal to g except, possibly, for the values assigned to x.

Groenendijk & Stokhof give a very elegant application of this approach to the treatment of anaphora. When interpreting the discourse *A man is walking in the park. He whistles*, it is possible to first construct a representation for the first sentence, and subsequently add a representation for the second sentence and conjoin this to the representation of the first:

(14) $(\exists x : [MAN(x)$ & $WITP(x)])$; $WHISTLE(x)$

The ";" in this formula represents a noncommutative *and*, whose semantics is defined as follows:

(15) $| | \phi;\psi | | = \{<g,h> |$ there is an assignment k such that
$<g,k> \in | |\phi| |$ and $<k,h> \in | |\psi| |\}$

The combination of (13) and (15) has the amazing effect that the free variable x in $WHISTLE(x)$ is bound by the existential quantifier, even though it is outside its scope. Intuitively, this is precisely how anaphora seem to work in natural language.

Whether this dynamic form of model-theoretic semantics can be applied successfully to other discourse phenomena than anaphora still remains to be seen. The underlying idea that "the meaning of a sentence does not lie in its truth conditions, but rather in the way it changes the (representation of the) information of the interpreter" is quite similar to the common starting point of discourse-semantic theories, which view the interpretation of multi-sentence discourse as a process that grows and updates representation structures (see Guenthner, 1988). It is also closely akin to the idea that the pragmatic meaning of an utterance consists in the way the semantic content changes the mental states of the participants involved.

The combination of these ideas, put forward in Bunt (1988a; 1988b; 1990), leads to a dynamic semantic/pragmatic interpretation theory not confined to the sentence level, along the following lines.

First, the representation structures grown incrementally according to discourse theories, are construed as *representations of the mental states* of the agents involved. These representations are highly complex structures, having components that consist of information relative to a certain propositional attitude and a certain agent (including multiple nestings and mutual beliefs). Second, the pragmatic part of the meaning of an utterance (its communicative function) is construed as identifying the components of the current representation structures that the utterance addresses. Third, and finally, the semantic part of the meaning of an utterance is construed as indicating how the information within the components, identified by the pragmatic meaning, is to change.

A perlocutionary dimension can be added to this framework as the way in which further changes occur in the contents of the components involved of the dynamic representations.

Such a framework comprises the common ideas of incremental, dynamic interpretation in currently developing theories of sentence and discourse semantics, and in AI in work on pragmatics and default reasoning. The further development of a unifying framework along these lines holds the perspective of a theory of natural language understanding and communication in which all the pieces fit together that are now often scattered over different partial theories. Bringing them together in one coherent theory may be expected to contribute to our understanding of language understanding as well as to the establishment of the foundations of language understanding systems.

REFERENCES

Allen, J. (1988). Natural language understanding. Benjamin/Cummings, Menlo Park, CA.

Allwood, J. (1976). Linguistic communication as action and cooperation. Gothenburg Monographs in Linguistics 1, Gothenburg University.

Alshawi, H. (1990). Resolving quasi-logical forms. Computational linguistics 16 (3), 133-134.

Alshawi, H., Moore, R.C., Moran, D.B. & Pulman, S.G. (1987). Research Program in Natural-language Processing, annual report. SRI International, Cambridge Computer Science Research Centre, July 1987.

Appelt, D. & Konolige, K. (1988). A practical nonmonotonic theory for reasoning about speech acts. SRI Technical Note 432, April 1988.

Barwise, J. & Perry, J. (1983). Situations and Attitudes. Bradford Books/MIT Press, London & Cambridge, Mass.

Beierle, C., Dörre, J., Pletat, U., Rollinger, C., Schmitt, P. and Studer, R. (1988). The knowledge representation language LLILOG. LILOG Report 41, IBM Deutschland, Stuttgart.

Beun, R.J. (1988). Declarative or interrogative, that is the question. Forthcoming in Journal of Pragmatics.

Beun, R.J. (1989). Declarative question acts. In : M.M. Taylor, F. Néel & D. Bouwhuis (eds.). The structure of multimodal dialogue. North-Holland, Amsterdam.

Black, W.J. et al. (1991). A pragmatics-based language understanding system. Forthcoming in Proceedings of ESPRIT '91 Conference, Brussels, November 1991.

Bouma, G., König & Uszkoreit, H. (1988). A flexible graph-unification formalism and its application to natural-language processing. IBM Journal of Research and Development 32 (2), 170 - 184.

Bronnenberg, W.J., Bunt, H.C., Landsbergen, S.P.J., Scha, R.J.H., Schoenmakers, W.J., & Utteren, E.P.C. van (1980). The question answering system PHLIQA1. In: L. Bolc (ed.) Natural language communication with computers. Hauser, Munich.

Bunt, H.C. (1984). The treatment of quantificational ambiguity in the TENDUM system. In Proceedings COLING'84, Stanford, CA.

Bunt, H.C. (1985). Mass terms and model-theoretic semantics. Cambridge University Press, Cambridge.

Bunt, H.C. (1986). Information dialogues as communicative action in relation to partner modelling and information processing. In : M.M. Taylor, F. Néel & D. Bouwhuis (eds.). The structure of multimodal dialogue. North-Holland, Amsterdam 1989.

Bunt, H.C. (1988a). On-line interpretation in speech understanding and dialogues systems. In: H. Niemann, M. Lang, & G. Sagerer (eds.) Recent advances in speech understanding and dialog systems. Springer, Berlin.

Bunt, H.C. (1988b). Towards a dynamic interpretation theory of utterances in dialogue. In: H. Bouma & B. Elsendoorn (eds.) Working models of human perception. Academic Press, New York.

Bunt, H.C. (1990). DIT–dynamic interpretation in text and dialogue. In: L. Kálmán & L. Pólos (eds.) Papers from the Second Symposium on Logic and Language. Akadémiai Kiadó, Budapest, 67-104.

Bunt, H.C., Beun, R.J., Dols, F.J.H., Linden, J.A. van der & Schwartzenberg, G.O. (1985). The TENDUM dialogue system and its theoretical basis. IPO Annual Progress Report 19, 105-113.

Calder, J., Klein, E., Moens, M. & Zeevat, H. (1987). Problems of dialogue parsing. ACORD Research Paper EUCCS/RP-1, Centre for Cognitive Science, University of Edinburgh.

Cohen, P.R. & Levesque, H.J. (1985). Speech Acts and Rationality. In: Proc. 23rd Annual Meeting of the ACL, pp. 49-59.

Colban, E. & Fenstad, J.E. (1987). Situations and prepositional phrases. In: Proceedings of ACL Europe Conference, Copenhagen.

Davidson, D. (1967). The logical form of action sentences. Reprinted in his Essays on Actions and Events. Clarendon Press, Oxford 1980.

Deemter, C.J. van, Brockhoff, G., Bunt, H.C., Meya, M., & de Vet, J.M. (1985). From TENDUM to SPICOS, or: how flexible is the TENDUM approach to question answering? IPO Annual Progress Report 20, 83 - 90.

Fenstad, J.E. (1988). Natural Language Systems. In: R.T. Nossum (ed.) Advanced Topics in Artificial Intelligence, Lecture Notes in Computer Science 345, Springer, Berlin.

Fenstad, J.E., Halvorsen, P.-K., Langholm, T. & van Benthem, J. (1987). Situations, Language and Logic, Reidel, Dordrecht.

Gazdar, G. (1981). Speech act assignment. In: A.K. Joshi, I. Sag & B.L. Webber (eds.) Elements of discourse understanding. Cambridge University Press, Cambridge, U.K.

Groenendijk, J.A.G. & Stokhof, M.B.J. (1987). Dynamic predicate logic. Unpublished paper, ITLI, University of Amsterdam.

Groenendijk, J.A.G. & Stokhof, M.B.J. (1988). Context and information in dynamic semantics. In: H. Bouma & B. Elsendoorn (eds.) Working models of human perception. Academic Press, New York.

Guenthner, F. (1989). Discourse: understanding in context. In: H. Schnelle & N.O. Bernsen Logic and Linguistics: Research directions in cognitive science: European perspectives, vol 2, Hove, U.K. Lawrence Erlbaum Associates Ltd.

Harel, D. (1984). Dynamic Logic. In: D. Gabbay & F. Guenthner (eds.) Handbook of Philosophical Logic, vol II. Reidel, Dordrecht.

Herzog, O. et al. (1986). LILOG - Linguistic and logic methods for the computational understanding of German. LILOG Report 1b, IBM Germany, Stuttgart.

Imlah, W.G. (1987). CML in LOQUI. Unpublished LOKI working paper BI-30, Hamburg University.

Kamp, H. (1981). A theory of truth and semantic interpretation. In: J. Groenendijk & M. Stokhof (eds.) Formal methods in the study of language. Mathematisch Centrum, Amsterdam.

Levinson, S.C. (1983). Pragmatics. Cambridge University Press, Cambridge, U.K.

Montague, R. (1974). Formal Philosophy. Yale University Press, New Haven.

Moore, R.C. (1980). Reasoning about knowledge and action. SRI Technical Note 151, SRI International, Menlo Park, Cal.

Perrault, C.R. (1988). An application of default logic to speech act theory. In : M.M. Taylor, F. Néel & D. Bouwhuis (eds.) The structure of multimodal dialogue. North-Holland, Amsterdam.

Pulman, S. (1987). Events and VP modifiers. In: Recent developments and applications of natural language understanding. Unicom, Uxbridge.

Scha, R.J.H. (1983). Logical foundations for question answering. Ph.D. thesis, University of Groningen.

Saebo, K.J. (1988). A cooperative yes-no query system featuring discourse particles. Proceedings of COLING'88, Budapest, 549 - 554.

Smolka, G. (1988). A Feature Logic with Subsorts. LILOG Report 33, IBM Germany, Stuttgart.

Thurmair, G. (1987). Semantic processing in speech understanding. In: H. Niemann, M. Lang, & G. Sagerer (eds.) Recent advances in speech understanding and dialog systems. Springer, Berlin.

Vestre, E. (1987). Representasjon av Direkte Spoersmaal. Cand. Scient. Thesis, University of Oslo.

Wachtel, T. (1987). Discourse structure in LOQUI. In: Recent developments and applications of natural language understanding. Unicom, Uxbridge.

Wahlster, W. (1988). Research trends in natural language systems. In Cognitive Science in Europe vol. 2. (forthcoming).

Winograd, T. (1972). Understanding Natural Language. Edinburgh University Press.

NOTES

1. The argument that follows is independent of the particular, somewhat naive meaning representation used here.

2. The computational perspective in fact adds a second criterion of adequacy: tractability (see section 3.1).

3. We return to this view in section 4, in the more specific form where meaning is equated with state-changing potential.

4. An interesting attempt to get around this difficulty can be found in Moore (1980), where possible-worlds variables are quantified over in an extensional first-order object language.

CHAPTER 3

Characterising Machine Learning Programs: A European Compilation

Yves Kodratoff
*CNRS & Université Paris Sud, LRI, Bldg 490, 91405
Orsay, France*

ABSTRACT

The first part of this paper is an introduction to the Machine Learning (ML) methodologies stemming from the Artificial Intelligence (AI) approach.

The second part describes a number of possible characteristic features of ML programs. We present a questionnaire as it was submitted to European researchers, then summarize their answers in tables, and conclude with a quick analysis of the answers. The questions we have been asking were chosen to mark the Artificial Intelligence (AI) approach to ML, as illustrated for instance in [Michalski et al. 1983, 1986, Kodratoff 1988]. It follows that some of the systems that perform learning essentially by adjusting coefficients may find it difficult to fit into our schemes. On the contrary, all AI oriented programs should fit in, unless we have made mistakes with our description.

Artificial as it looks, this effort of characterization addresses two deep concerns. The first one is a clarification of the terms used, and sometimes abused, in the ML community. The second is to promote communication in the sub-field. In most cases, learning systems do learn, but it is neither clear what nor how they learn. Setting bench marks for ML is a necessity the community should be more aware of.

Some of the systems known as knowledge acquisition systems perform some learning as well. The last section is devoted to this kind of program.

1. INTRODUCTION

This introduction will give an overall description of the topics in ML research, and a description of what is the AI approach to ML.

Figure 3.1 is a description of ML that comes from the introduction of the third volume in the series *ML: An AI Approach* [Michalski et al. 1983, 1986, Kodratoff and Michalski 1990].

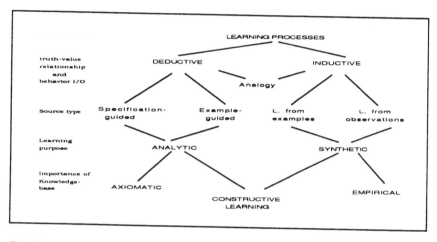

Fig. 3.1.   Michalski's general description of ML.

This description states that constructive learning, the goal of which is the creation of new concepts and descriptors, cannot be achieved without merging inductive learning (that will generate hypotheses about the new knowledge) and deductive learning (that is necessary to validate the hypotheses). This view of constructive learning, which the author fully agrees with, is however far from being universally accepted. This is why we have added question 6.11 (see section 6) in our questionnaire.

This drawing states that induction by itself does not reduce to empirical learning. In empirical learning, very little use is made of the background knowledge. In many cases, empirical learning reduces to keeping what is common between the examples, or to coefficient optimising. We want to insist here on the fact that knowledge intensive "constructive induction" is also possible. Section 1.2 gives a detailed example of such knowledge intensive inductive learning.

The taxonomies above give a fairly accurate description of the different ML topics, and of their interrelationships. It shows a striking symmetry between deductive and inductive learning. Most of the early

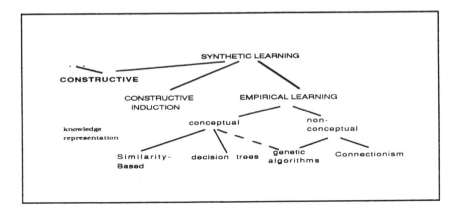

Fig. 3.2.   Details on synthetic learning.

research has been on inductive learning, in the USA as well as in Europe. The present symmetry reflects the so-called "Explanation-Based learning (EBL) revolution" that took place during the last few years mainly in the USA, under the influence of Gerry DeJong and Tom Mitchell. By consulting the tables in which the European systems are described, one notices that EBL has become popular in Europe as well.

We have highlighted the main implications of this taxonomy, the rest is quite self-explanatory, if the terms used in it are not ambiguous. There are still discussions about their exact meaning. The glossary in the appendix describes some of the most important choices that have been made in this paper.

Section 9 will provide some information about a further field known as Knowledge Acquisition (KA), defined in the glossary. We felt it necessary to include in this compilation those KA systems that use ML

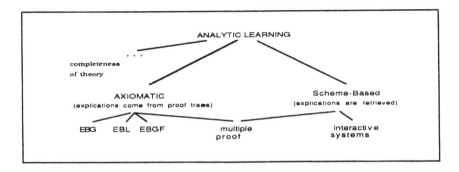

Fig. 3.3.   Details on analytic learning.

because they are actually so few of them, and because this tendency is well represented in Europe.

The AI approach to ML relies on several concepts (induction, abduction, generalization) that are often misunderstood. We shall therefore first give a precise definition of them.

1.1. Induction, Deduction

Let us first as briefly as possible recall what induction and deduction are, more details are given in [Kodratoff 1988].

Deduction is the process by which one **infers while preserving the truth** of the data already stored. Applying the well-known modus ponens rule is truth-preserving. For instance, suppose that a theorem tells us that all green leaves are young leaves. On seeing green leaves, we can deduce that they are young.

Induction (Note 1.) is the process by which one **infers while preserving the falsity** contained in the data. Michalski generalization rules [Michalski 1984] are falsity preserving. Actually, two kinds of false statements must be preserved: Falsity can originate from the examples or from the background knowledge. When falsity originates from the examples, this simply means that the formula obtained by induction from the examples must not take the value true on examples that are false. As a first instance, consider a scene that contains green leaves and red leaves. From this scene, and if we do not use our knowledge of colours, we can induce $\supset x$ [leaf(x) & green(x)]. As a second and reverse instance, we cannot induce $\supseteq x$ [leaf(x) & green(x)] when we are looking at a scene of which we explicitly know that it contains green and not green leaves, because we explicitly know that [leaf(that_leaf) & green(that_leaf)] is false in the scene we are working with. In other words, suppose that "that_leaf" is red, and not green, then $\supset x$ [leaf(x) & green(x)] does not preserve the falsity of [leaf(that_leaf) & green(that_leaf)]. These two examples illustrate also the extent to which induction without background knowledge may be crude.

This is why we must consider that falsity can also originate from the background knowledge. For instance, suppose that we are looking at a scene with green leaves only, and that the background knowledge tells us "green" implies "not-red", i.e. $\supset x$ [green(x) $\Rightarrow \neg$red(x)], which we will use under the naive form $\supseteq x$ [[green(x) = TRUE & red(x) = FALSE] $\vee$ [green(x) = FALSE & red(x) = TRUE]]. Using this theorem, the scene with n green leaves (for which "green" happens to be true) can be described by the formula [[leaf(that_leaf1) & green(that_leaf1)] & [leaf(that_leaf2) & green(that_leaf2)] & ... & [leaf(that_leafn) & green(that_leafn)]] = TRUE & [[leaf(that_leaf1) & red(that_leaf1)] &

[leaf(that_leaf2) & red(that_leaf2)] & ... & [leaf(that_leafn) & red(that_leafn)]] = FALSE. While inducing, we must keep everything which is false. For instance, we can induce that we are seeing green things because no background knowledge links the colour green to some falsity about leaves. Conversely, we cannot induce that we are seeing red leaves, which would not preserve the falsity of the second part of the formula.

1.2. Knowledge Intensive Induction

Many authors link deductive learning with knowledge intensive approaches (which they actually are in existing implementations), and inductive learning with knowledge poor approaches (there exist implementations of induction that are knowledge intensive as well, for instance [Michalski 1983, Kodratoff et al. 1984, Kodratoff & Ganascia 1986]). It is clear that "pure induction" is knowledge poor. The point is that pure induction would mean induction from raw data, and raw data does not exist in reality since the world representation we put into the machine, as raw as it is, is bound to already be considerably interpreted. A clear definition of abduction will help us to define induction better, and to again highlight the role of background knowledge in the global inductive process.

1.2.1 - What is abduction?

Technically, abduction is the process which "reverses" modus ponens. As well-known, modus ponens is an inference process by which, knowing A, and $A \Rightarrow B$, one infers B. Abduction is the inference process by which knowing B, and $A \Rightarrow B$, one infers A. Of course, this inference may introduce contradictions in a theory since it might happen that B is True, A is false, therefore the implication $A \Rightarrow B$ is trivially True, which shows that abduction is not truth-preserving. In practice, abduction is mainly used during the process by which we are able to complete a proof that just failed, by creating hypotheses that, if they were valid, would indeed lead to a complete proof.

Let us illustrate the practical use of abduction by the following example drawn from [DeJong and Mooney 1986]. In this example, our aim is learning a definition of the concept of suicide Kill(X,X). Suppose then that we have been learning so far that suicide requires a weapon, as expressed by the rule

KILL(X,X):- DEPRESSED(X), BUY(X,Y), WEAPON(Y).

Suppose now that we know Mary commits suicide. Suppose that we have the following knowledge about her:

DEPRESSED(MARY).
BUY(MARY,OBJ1).
SLEEPING-PILLS(OBJ1).
PRICE(OBJ1, 20).
BUY(MARY, OBJ2).
BOOKS(OBJ2).
. . .

where OBJ1 and OBJ2 are constants.

We will be unable to prove KILL(MARY, MARY) because she has no weapon. Cox [Cox and Pietrzykowski, 1986] presents an algorithm that, in such a situation where WEAPON(Y) fails, would automatically look for "weapons" and catch the first objects that have been satisfactory so far (here, those that have satisfied BUY(X,Y)), therefore abduct WEAPON(sleeping-pill) or WEAPON(book) depending on which is met first. In the absence of any other kind of knowledge, it is clear that anything that has been bought might be taken as a weapon. We [Duval and Kodratoff 1989] support the fact that another abduction is also possible, namely by adding new rules in which the objects that have been satisfying BUY(X,Y) are added into the clause, without hypothesizing that they are "weapons". Here, one could abduct as well

KILL(X,X):- DEPRESSED(X), BUY(X,Y), SLEEPING-PILLS(Y)

or

KILL(X,X):- DEPRESSED(X), BUY(X,Y), BOOKS(Y).

All this shows that even in very constrained environments, many abductions are possible, all of them solving the problem of completing the failed proof. Choosing which is the best one depends clearly on the background knowledge at hand. This is why we claim that even though an isolated abduction is knowledge poor, the whole abductive process is very much knowledge intensive.

1.3. Generalization/Particularization

Tom Mitchell's version spaces [Mitchell 1982, Genesereth and Nilsson 1985, Kodratoff 1988] maintain the generalization state of an operator. They give the exact generalization state in which a descriptor used by an operator must be kept in order to optimise the problem solving efficiency. Given a set of positive and negative examples, their "version space" is the set of the consistent formulas, i.e. the set of formulas that are both complete (they recognize all the positive examples) and coherent (they recognize none of the negative examples).

When building the version space of an operator, positive examples are used "bottom-up", that is to say, by finding a formula which is more general than any of the examples, and such that each example is an instantiation of this formula (also, none of the negative examples should be an instantiation of it, but this is not the point here). Conversely, the negative examples are mainly used "top-down", since they are used to particularize a formula that excludes all negative examples. This should convince people that generalization (from positive examples) and particularization (from negative examples) are the two facets of the same process, known under the name of "generalization" because earlier works did not recognize enough the importance of negative examples, and the symmetry between positive and negative examples [Nicolas 1988].

There are other approaches than those inspired by the version spaces. Michalski's [Michalski and Stepp 1983, Michalski 1984] work is typical of those that progressively generalize from examples (his algorithms perform bottom up generalization). These algorithms cluster examples together, and simultaneously find a function that recognizes these clustered examples and reject all counter-examples. The progressive growth of these clusters generates more and more general recognition functions. On the contrary, all the ID3 family inspired from Quinlan [Quinlan 1983] start from a set that contains all the examples, and progressively apply descriptors that cut it into sub-sets, according to the value of the descriptor. For instance, if we start with a set containing persons with all types of eye colours, we shall form as many sub-sets as there are eye colours, each containing a single colour. Therefore, these algorithms progressively particularize, they perform top-down generalization, as some put it.

Generalization is always an inductive process. It is therefore very wrong to confuse particularization and deduction. To make this precise, let us describe in a similar way generalization and particularization. When generalizing (sometimes called bottom up generalization), the problem is inducing the common features existing between several examples. These common features characterize the concept they are all instances of. When particularizing (sometimes called top-down generalization), the problem is inducing the features that make the difference between several examples. Each example is characterized by the features that are not shown by the others.

1.4. What is the AI Approach to ML

We present here five features characterizing AI, that make the AI approach to ML original. When they are used in a very strict way, that is to say when all of them are requested to be fulfilled, one defines something like the core of AI. A softer definition of AI, that will include all the existing AI systems, will be to request that some of the features given below are partly fulfilled.

First, as opposed to the "intelligent machine" definition [Barr and Feigenbaum 1981], AI refuses to built intelligent black boxes. For instance, Roger Schank points out [Schank 1986] that the chess playing machines presently available on the market do not use AI techniques even though they start being real good players. The reasons why they play so well are available to their designers only, not to their users, and therefore they cannot be considered as AI. Anyhow, it would be somewhat strange to attribute to AI the success of these chess machines since they result from Operation Research, not from AI. We do not claim here that temporary stoppage of communication is impossible in AI, we rather point out that a definitive loss of any communication abilities is contrary to the spirit of AI. AI systems are open to their users who must understand them.

Second, AI tends to avoid using implicit knowledge. A clear illustration of this assertion is the marked preference AI shows for declarative knowledge representation, in which knowledge and the way to use it are clearly differentiated. AI tends to avoid procedural representations where knowledge and its use are merged. For instance, AI favoured the birth of rather strongly declarative languages like LISP and PROLOG. Another well-known example is Expert Systems (ES). An ES is nothing but a kind of decision tree, except that the tree has been flattened in a declarative form by writing out each rule that leads to a decision, without specifying in advance in which order they must be called. This generates the so-called "conflict resolution" problem, which is the price that must be paid for explicit declarative knowledge. In an AI system, knowledge should be stated explicitly, preferably in a declarative way. As little as possible procedural knowledge is expressed in a programming language. Similarly, fancy ad hoc codings are very much in opposition with the spirit of AI.

Third, as opposed to what the early workers believed, a system cannot fulfil the above two requirements without already carrying a huge amount of preliminary knowledge. At least, this knowledge must carry the user's understanding of the problem. This is illustrated by the importance given to knowledge bases in AI. Knowledge is a kind of data: We do not speak of data bases because our bases have to be declarative and understandable by their users. An AI system is supposed to make heavy use of background knowledge provided by the user in a knowledge base.

Fourth, in accordance with the need for user understandability, an AI system has to speak its user's language, instead of speaking a "language of AI", which, incidentally, does not exist. For instance, statistics does describe interesting natural phenomena, and does it in terms of "mean squares distances" etc ..., which is its own language, instead of expressing the knowledge in its user's language, as AI attempts (and still partly fails) to do.

Fifth, we believe (even if this is far from being accepted by all our colleagues) that AI is concerned with the definition, analysis, measurement, and comparison of explanations relative to an automatically executed reasoning session, and provided in the user's language. Our main argument follows from the recognition that explicability fulfils the needs expressed above since the concern for understandability is perfectly met by a system able to provide its user with explanations. Moreover, a huge amount of background knowledge is necessary to be able to find relevant explanations. This is perfectly illustrated by explanation-based learning, a new way to deal with automatic knowledge acquisition developed during the last few years.

2. EXPLANATION-BASED LEARNING (EBL) IN STRONG THEORY DOMAINS

The most rigorous EBL learns generalizations only, this is why it has been called Explanation-Based Generalization (EBG).

2.1. EBG: An Intuitive Presentation

EBG is given a complete theory of the domain in which the learning takes place and a training instance of the concept to be learned. Since the system knows the complete theory, it knows a definition of the concept to be learned. The goal of the learning process is to learn a new "better" definition of this concept. To achieve this goal, EBG works as follows. Using its knowledge of the domain theory, it proves that the training instance is actually an instance of the concept learned. The proof trace is then pruned by using an operationality criterion. This criterion specifies which are the descriptors that are operational, i.e. which may be used in a rule. It is given as a list that contain all operational descriptors. This amounts to specifying which part of the proof is considered as being of interest. Notice the importance of this criterion on which relies all the efficiency of the process. The pruned proof trace is then generalized into an explanation structure, which contains the parts of the original clauses that have been used. As noted by DeJong and Mooney (1986), this amounts to keeping a part of the unifications needed to achieve the proof. A variablized version of the training instance is then regressed [Waldinger 1977] through the explanation structure. The result of this regression is a new, more operational definition of the concept.

In order to illustrate EBG, we shall use here a PROLOG version of the "SAFE-TO-STACK" example, taken from [Mitchell and al. 1986]. Several versions of EBG have been implemented as for instance in [Kedar-Cabelli and McCarty 1987, Puget 1987, Siqueira and Puget 1988]. The "SAFE-TO-STACK example shows how to learn a more efficient rule to stack given:

a definition of a theory of stacking and
an example of a particular box "box1" stacked on a particular table "endtable1".

In this particular example it happens that one needs to know the default value of the type "table". The first kind of information is the one relative to the specific box, "box1", and the specific table "endtable1".

C1 ON(box1, endtable1).
C2 COLOR(box1, RED).
C3 COLOR(endtable1, BLUE).
C4 VOLUME(box1, 10).
C5 DENSITY(box1, 1).
C6 FRAGILE(endtable1).
C7 OWNER(endtable1, CLYDE).
C8 OWNER(box1, BONNIE).
C9 ISA(endtable1, table).

The second kind of information is the one relative to the background knowledge about stacking things, also called the domain theory.

C10 SAFE-TO-STACK(X, Y):- NOT FRAGILE(Y)
C11 SAFE-TO-STACK(X, Y):- LIGHTER(X, Y)
C12 WEIGHT(X, W):- VOLUME(X, V), DENSITY(X, D), W is V*D
C13 WEIGHT(table, 50):-
C14 LIGHTER(X, Y):- WEIGHT(X, W1), WEIGHT(Y, W2), LESS(W1, W2)
C15 LESS(X, Y):- X < Y

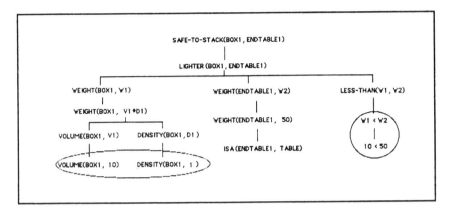

Fig. 3.4.   Tree of the proof that "box 1" can be stacked on "endtable 1".

The third kind of information is the one expressing that one can safely stack "box1" on "endtable1". Since we want to prove that by a PROLOG refutation procedure, this knowledge will be given as question to the PROLOG interpreter. It reads:

C16   :- SAFE-TO-STACK(box1, endtable1)

The proof proceeds as shown by the following trace. This trace is provided by most PROLOG interpreters, though in less readable form.

In this trace the descriptors we have chosen to call non-operational have been encircled. By pruning the non-operational descriptors and generalizing, we obtain the following explanation structure.

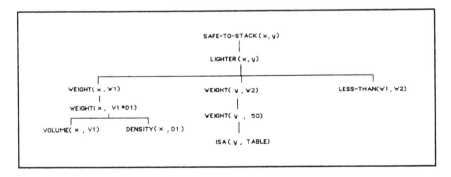

Fig. 3.5.   Pruned tree of the proof. Non-operational descriptors have been removed.

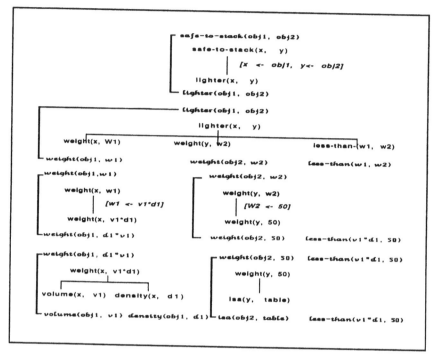

Fig. 3.6.   Goal regression through the pruned tree of the proof.

EBG then regresses a general goal through the explanation structure, and propagates the obtained instantiations through the whole set of regressed sub-goals. The final set of sub-goals becomes the new conditions for the achievement of the general goal. The figure below shows the explanation structure for SAFE-TO-STACK.

Around each part of the explanation structure are the goals and their regressions indicated in Venice font. At each level of regression, the substitutions are transmitted to all the sub-goals. The last line of sub-goals is the regression of "SAFE-TO-STACK(OBJ1, OBJ2)" through the explanation structure. The rule learned is therefore

SAFE-TO-STACK(OBJ1,   OBJ2):-   VOLUME(OBJ1,   V1), DENSITY(OBJ1, D1), LESS-THAN(V1 * D1, 50), ISA(OBJ2, table)

where V1, D1 are variables, and where the "50" comes from C13.

2.2. EBL and the Refinement of Strong Theories

For the management of incomplete theories, the central idea is to complete the failure proofs by an abduction mechanism [Duval and

Kodratoff 1989]. This abduction process is guided by analogical reasoning about explanations and enables the discovery of new rules that refine the domain theory.

Remember section 1.2.1 in which we considered that some training instances led us to learn:

KILL(X,X):- DEPRESSED(X), BUY(X,Y), WEAPON(Y).

Recall also the story of Mary's suicide, which comes with the information:

DEPRESSED(MARY).
BUY(MARY,OBJ1).
SLEEPING-PILLS(OBJ1).
ISA(sleeping-pill, OBJ1).
PRICE(OBJ1, 20).
BUY(MARY, OBJ2).
BOOKS(OBJ2).

We then proposed to explain Mary's suicide by adding

KILL(X,X):- DEPRESSED(X), BUY(X,Y), SLEEPING-PILLS(Y)

or

KILL(X,X):- DEPRESSED(X), BUY(X,Y), BOOKS(Y)

to the knowledge base. These two rules have been built because, when applied to this story, they lead to a resolution tree which is analogous to the one obtained to explain the earlier stories. In that way, the theory is completed since we are now able to prove Mary's suicide as well.

We are still left with the choice between the two possible clauses. One way to choose is considering theories of different granularities that support each other. By granularity of a theory, we mean the level of detail it copes with. For instance, a sociology of a suicide does not take into account details about the chemical reactions that result in the death process. Even though social details can be of much subtlety, we say that, in this context, molecular chemistry is of finer grain, and will be discarded in a first step. Suppose that our theory (theory1) has been so far of granularity 1, and that we have also a theory (theory2) of granularity 2 in which the "deeper" properties of the descriptors of theory1 are given. Theory2 will contain knowledge about the use of sleeping pills, books, and other objects that Mary may have been buying.

Theory2 will have also more detailed knowledge about the way killing is performed, for example about the potential danger of all weapons. Thus, one should be able to prove that the potential danger of a book is of pure psychological nature, while the potential danger of an excess of sleeping pills is also physical. One could then come back to theory1 with an explanation of why the sleeping pills, provided they are used in excess, are suitable tools for suicide. Notice however that all this reasoning is done with uncertain arguments. For instance, eating books in excess will certainly also cause death, it is simply less probable that one ingests books in a physical manner. Supposing these kinds of difficulties are taken care of, theory 2 can help revising theory 1 by adding:

KILL(X,X):- DEPRESSED(X), BUY(X,Y), SLEEPING-PILLS(Y), EXCESS-INGEST(X,Y)

If one is concerned only by choosing among the possible abductions, then one could also put aside all the stories unaccounted for by:

KILL(X,X):- DEPRESSED(X), BUY(X,Y), WEAPON(Y)

and, after many of them are found, cluster those that present common features. If the stories are varied enough then "sleeping pills" should be the only common feature among several examples.

2.3. Storing Chunks of Knowledge

After a problem solving session the trace, whole or partial, of the solution obtained can be kept. Since such solutions are then stored in the knowledge base, it can be used to obtain immediate solution in further applications. At this stage, the problem consists of keeping only the most interesting chunks.

In practice, many other problems occur when the chunks are simply piled together, and that the system starts crumbling under its own weight. One then has to recognize when two chunks are equivalent or can be generalized into a more efficient one, and to organize the chunks into "chunks of chunks" which amounts to using meta-knowledge to sort them.

The system SOAR [Laird et al. 1986, 1987] which implements some of these ideas is currently under test in an industrial environment and one should soon learn about its achievements.

3. EXPLANATION-BASED LEARNING IN WEAK THEORY DOMAINS

EBL assumes that one has a strong theory of the domain. This is not always the case, and AI developed methods adapted to weak deep theories, where by "weak" we mean: incomplete and/or incoherent and/or intractable. We shall describe here three different approaches for extracting explanations from weak theories. One is illustrated by the system PROTOS [Bareiss, Porter, and Wier 1990], the other one by the system DISCIPLE [Kodratoff and Tecuci 1987a, 1987b], and the third one by Schank's theory of XPs [Schank 1987, Schank and Kass, 1990].

3.1. Definition of an Explanation for PROTOS

In PROTOS, the production rules are associated with elementary explanations, that describe what kind of relation takes place between the antecedent and the consequent of each production rule. There are six such elementary explanations. Among them, one finds the classical ISA (generality relationship), and PART-OF (part to whole mappings), but also ENABLES indicating that a feature enables the function of an object (e.g., wings enable flight).

An explanation is a combination of such elementary explanations as given by a path in the semantic net of the objects.

For example, one can say that the relation between engine and car is PART-OF. But a better explanation can be provided by considering a less direct path: the engine ENABLES the movement which is the FUNCTION of vehicles and car ISA vehicle. This knowledge is part of the training provided by the teacher to underline the explanations that are significant.

In the above example, suppose the system is taught that the second path provides a good explanation, and not the first. Then, in the future, when trying to detect whether some unknown x is an engine, it will not try to prove that x is a part of an engine but will try to see if x can enable the movement of some vehicle.

In PROTOS a good explanation is a path in the semantic net of the objects, the path is provided by the user.

3.2. Definition of an Explanation for DISCIPLE

DISCIPLE is an interactive Learning Apprentice System for weak theory fields. In DISCIPLE, the learning process starts with an example of a problem solving step as, for instance, the following rule for the construction of loudspeakers.

Example 1
ATTACH sectors ON chassis-membrane-assembly ⊢—
 APPLY mowicoll ON sectors
 PRESS sectors ON chassis-membrane-assembly

where A ⊢—B means: "in order to achieve action A, perform actions B", and where "mowicoll" is a kind of glue, and "sectors" and "chassis-membrane-assembly" are parts of a loudspeaker. This is interpreted as an example of a general rule indicating a way of performing the ATTACH action:

General Rule1
ATTACH x ON y⊢—
 APPLY z ON x
 PRESS x ON y

Being completely instantiated, Example 1 implicitly contains the proof of its validity in the supposedly known properties of sectors, chassis-membrane-assembly, and mowicoll. On the contrary, since we have no knowledge of the possible properties of the variables found into General Rule1, it does not contain any implicit explanation.

Nevertheless, it is essential to obtain this knowledge, because it tells us when it is valid to perform an ATTACH operation as a sequence of APPLY and PRESS. In order to achieve this goal, the system over-generalizes the examples (by transforming Example 1 in General Rule1, for instance) and applies this over-generalization to the data base of the user in order to find instances of the over-generalization. These instances are proposed to the user as tentative rules. When the user validates such a system-produced rule, we say that the system is provided with a positive example of its over-generalization. When the user rejects such a rule, we speak of a negative example. This set of positive and negative examples is used by a classical generalizer such as AGAPE [Kodratoff et al. 1984] to produce a more refined generalization. Finally, from these interactions with the user, DISCIPLE may find what is the correct generalization of the proofs of the examples. This gives a correct set of conditions for the application of the generalized rule, and hence a good explanation.

For instance, the final rule DISCIPLE will learn, from Example 1 and further interactions with the user, is the following.

General Rule2
IF
  z ISA adhesive
  z TYPE pure
  z GLUES x
  z GLUES y
THEN
  ATTACH x ON y ⊢
    APPLY z ON x
    PRESS x ON y

Since the structure of the rule is precisely the structure of the example, what is to be learned is just the explanation of the rule.

An explanation in DISCIPLE is a set of conditions that authorizes the application of a general rule. For instance, the conditions of General Rule2 are an explanation of General Rule1. It is always obtained as the generalization of the particular proofs of the example rules provided to the system.

The definition of an explanation is therefore very similar to the one of section 2, the difference is in the generalization process that is used. DISCIPLE is just in the middle between strong-theory-EBL and PROTOS. In strong-theory-EBL the explanations are provided by the system to the user, in PROTOS they are provided by the user to the system, in DISCIPLE they are obtained as the final result of a user consultation with the system.

3.3. Pre-stored Explanations: Schank's XPs

Roger Schank has recently proposed his own vision of Machine Learning and AI, of which creativity is the central theme [Schank 1987]. Nevertheless, in this theory, the essential creative information is constituted by a set of Explanation Patterns (XPs). These XPs are pre-stored explanations that can be applied in a variety of situations. Scripts [Schank and Abelson 1977] are classical in AI, they contain pre-stored information about standard situations. Let us define an XP as being a script for an explanation.

An XP is made of four parts:

an index that allows one to get at the XP;

a set of states of the world under which the XP can expect to be activated;

a scenario which is a causal chain of states and events;

the resultant state following from the scenario application.

For instance, in the case of the unexpected death of a promising racehorse named Swale, Schank [Schank 1986] proposes several XPs containing the following scenarios:

"Early death comes from being malnourished as a youth"

"High living brings early death"

"An inactive mind can cause the body to suffer"

"High pressure jobs cause heart attacks"

...

The first scenario leads to ask whether Swale underwent bad treatments during its youth, etc ...

From the perspective of PROTOS and DISCIPLE the issue could be viewed as obtaining of new XPs, in which case, the explanations are pre-recorded and the problems are:

the retrieval of the relevant ones from a certainly huge number;

the application through some analogical instantiation process to the particular case under consideration;

their modifications to fit into unexpected situations (tweaking them, as Schank puts it).

In practice, one may assume some XPs will have to be provided manually, but also that no system will be able to run a real life problem without a mechanism to generate its own explanations.

4. ANALOGICAL REASONING

First, let us briefly recall what analogical reasoning is. For more details, see for instance [Chouraqui 1985], [Davies and Russell, 1987] or [Kodratoff 1988, 1990a, 1990b]. Let us suppose that we dispose of a piece of information, called the source S (which can be viewed as a knowledge base) and that this information can be put into the form of a doublet (A, B) in which B depends (often causally) on A. Let us suppose now that another piece of information, called the target T, can be put into the same form (A', B') with the requirement that there exists some resemblance between A and A'.

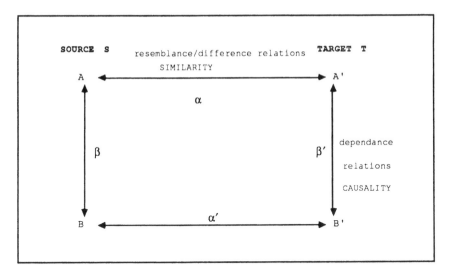

Fig. 3.7.   The general scheme of analogy.

This scheme can be used in different ways to define analogical reasoning. In most cases, we consider that we know more or less precisely, A, B, A', α and β, and that we try to invent or justify B'. Let us now describe two very different examples of analogy that will show how this scheme may be actually applied in a variety of ways.

In [Winston 1982], one is given a text A (e.g., Macbeth's story), an exercise A' (e.g., a short story between two other individuals, say, Tom and Jane) , and a question about A', B' (e.g., Is Tom likely to be evil?). The analogy problem is to answer B' (and to justify that answer), by considering the causal relationships in A. The similarity between A and A' is defined as their partial matching, i.e., α is the partial matching of A and A' (e.g., Tom and Jane are married, Tom is weak, and Jane is greedy. This matches partially Macbeth's story, since the same links hold between Macbeth and Lady-Macbeth). The part of A that matches A' is B (e.g., Macbeth and Lady-Macbeth are married, Macbeth is weak, and Lady-Macbeth is greedy), and the causality links inside B form the β of our scheme. Suppose that in Macbeth's story one assumes that his weakness and his wife's greediness is cause of his evil. One then postulates that the same causality links hold between A' and B', i.e., β = β'. If these causality links are relative to B' question (e.g., here, the causality links are relative to an explanation of Macbeth's evil), then one answers "yes' to B', using these causality links (e.g., Yes, Tom may be evil, because of his weakness, and his wife's greediness).

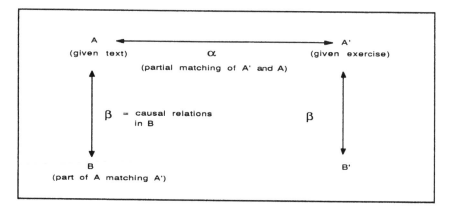

Fig. 3.8.   Winston's scheme of analogy.

In conclusion, one can summarize Winston's approach by the following scheme.

In [Carbonell 1983, 1986], one is given a completely solved problem that can be decomposed into its means A, and its goals B (e.g., the means may be "abducting a rich person's child", and the goals be "kidnapper becomes rich"), a plan that solves how achieving goal B by means A, which here can be viewed as the dependency β linking A and B. One is also given another problem, i.e., goals B' and means A' (e.g., A' may be: abducting a famous politician, and B' may be: kidnappers (terrorists) advertise their political cause). The analogical problem is then to use the plan β in order to invent a new plan β' that will achieve B' by A' (e.g.,

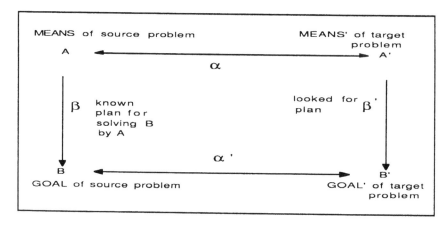

Fig. 3.9.   Carbonell's scheme of analogy.

how a terrorist can get publicity by abducting a famous politician?). In short, Carbonell gives A, B, A', B', β, and asks analogy to provide reasonable β'. He describes sets of possible transforms that can be applied to β in order to generate candidate β'.

In conclusion, one can summarize Carbonell's approach by the scheme in Fig. 3.9.

Notice that there exist several other approaches to analogy that we did not study here. The interested reader should refer to [Michalski et al. 1983, 1986, Priedetis 1988, Kodratoff 1988, 1990a, 1990b].

5. INDUCTIVE LEARNING: ID3, INDUCE, STRUCTURAL MATCHING

We present here three induction methods ordered by the amount of background knowledge they rely on. ID3 is typically knowledge poor. Nevertheless, all the work done on ID3 from the original version [Quinlan 1983] to its recent improvements (see for instance [Bratko and Lavrac 1987]) are centered on understandability rather than on efficiency, which, by the way, gives to these works their deep originality as well as their belonging to AI. Surprisingly enough, it seems that most improvements in understandability have been followed by efficiency increase as well [Bratko 1988].

5.1. ID3

5.1.1. Stating the Problem
The aim of this method is as follows. Given a set of descriptors, of examples and of concepts the examples belong to, find the most efficient way to classify all the examples under the concept they belong to, i.e. find the most efficient way to "recognize" the examples. The method relies on information theory; it measures the amount of information associated with each descriptor, and chooses the most informative one. By applying descriptors in succession, a decision tree is built that will recognize the examples. As opposed to similar numerical techniques,

CONCEPT A			CONCEPT B		
size	nationality	family	size	nationality	family
small	German	single	small	Italian	single
large	French	single	large	German	married
large	German	single	large	Italian	single
			large	Italian	married
			small	German	married

ID3 preserves the understandability of its results by avoiding introducing linear combinations of descriptors.

Example. Suppose that we start with 2 concepts (A and B) described by 3 descriptors (size, nationality, family) that can take the values (small, large) for the descriptor size, (French, German, Italian) for nationality, and (married, single) for family.

Suppose that the concepts are illustrated by the examples in table 1.

ID3 will find an optimized decision tree to recognize concept A from concept B.

5.1.2. Numerical Optimization

For each descriptor that has not yet been used, the disorder left after applying a descriptor in the decision tree is computed. The one that leaves the least disorder is chosen as the next node of the tree. The process stops when each leaf of the tree contains examples of one concept only. The disorder is computed from the classical entropy formula

$$E = - \Sigma f_s \log_2 f_s$$

where f_s is the relative frequency of class s.

In our example the initial disorder is given by

$$E0 = -(3/8 \log_2 3/8 + 5/8 \log_2 5/8) = 0.954$$

Let us now compute the information gain of applying each descriptor. For instance, the information value of *nationality* is obtained by analyzing the result of classifying the examples by *nationality* and computing the disorder after applying nationality. This leads to the results shown in Fig. 3.10.

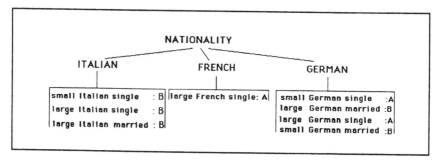

Fig. 3.10.   Result of the application of descriptor "NATIONALITY".

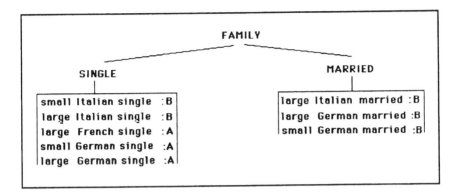

Fig. 3.11.   Result of the application of descriptor "FAMILY".

The disorder in the cluster *Italian* is: -(3/3 $\log_2$ 3/3) = 0, in the cluster *French*, it is -(1/1 $\log_2$ 1/1) = 0, and in the cluster German: -(2/4 $\log_2$ 2/4

+ 2/4 $\log_2$ 2/4) = 1. Therefore, the disorder left after applying *nationality* is

E1 = (3/8 * 0) + (1/8 * 0) + (4/8 * 1) = 0.5

The gain in information is E0 - E1 = 0.454.

The same computation is done for, say, *family*. Applying *family*, we get Fig 3.11.

The disorder after applying *family*, in the cluster single: -(2/5 $\log_2$ 2/5 + 3/5 $\log_2$ 3/5) = 0.971, and in the cluster *married*: -(3/3 $\log_2$ 3/3) = 0. The disorder left after applying *family* is therefore:

E'1 = 5/8 * 0.971 + 3/8 * 0 = 0.607

and the associated gain in information is E0 - E'1 = 0.347. The gain is less than when applying *nationality*, which should therefore be preferred to *family*.

In order to complete our example and show a decision tree, let us now compute the information value of *nationality* followed by *family*.

Applying them in order leads to the results shown in Fig. 3.12 in which all leaves contain examples of a single concept. This is therefore a complete decision tree allowing us to decide whether an example belongs to concept A or to concept B.

From the initial examples, we knew that applying the three descriptors in any order was indeed a decision procedure, we have now

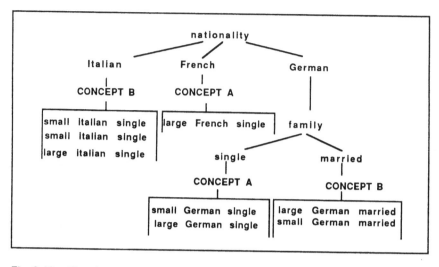

Fig. 3.12. Result of the application of descriptors "NATIONALITY" followed by "FAMILY".

learned that applying *nationality* and *family* is enough, and that we optimize the search speed by applying them in that order.

5.1.3. Improvements on ID3
All improvements performed are relative to a better understandability of the results. For instance, in some cases it seemed that preferring binary trees branching on their right only would improve understandability [Arbab and Michie 1985]. Other notable improvements are:
- pruning the tree (see [Bratko and Lavrac 1987]),
- transforming the tree into decision rules [Corlett 1983]
- using background knowledge [Kodratoff et al. 1987, Nunez 1988]

5.2. INDUCE [Michalski 1983]

5.2.1. Methodology
INDUCE builds a function R that recognizes positive examples and rejects negative ones, i.e. a complete and coherent function. In order to achieve this goal, it starts by building the most general recognition function R_i which recognizes the example e_i and rejects the set of examples {ej}. Details on how to build such a function will be given later.

Let POS be the set of positive examples, and NEG be the set of negative examples, let e_i be the current member of POS. The algorithm to compute R is as follows:

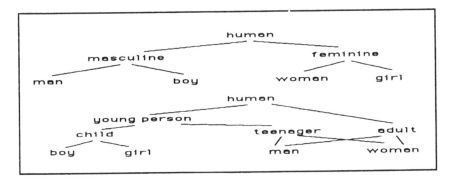

Fig. 3.13. "Taxonomies" of generality between the descriptors.

First, as we said, suppose that we have been able to compute R_i, the recognition function that recognizes e_i (and possibly some others of POS), and rejects all instances of NEG. If it happens that R_i recognizes all of POS, we have obtained a complete and consistent recognition function, $R = R_i$. If not, choose another example e_k, and compute R_k. Then, $R_i \vee R_k$ (where $\vee$ is the logical disjunction) will improve on e_k, and it is consistent by construction. Continue as long as $R_i \vee ... \vee R_n$ is not complete.

Example. Suppose we have ten examples described by means of four descriptors and their values, as shown in table 2.

The background knowledge is expressed by the properties of the descriptors. They can be linear like X1, the descriptor describing the values of weight. This will be expressed by the formula {light< medium-weight < heavy}.

Some descriptors can also be structured like X2 and X3.

The background knowledge relative to X2 may be given by the taxonomies of generality shown in Fig. 3.13.

Table 2.

	X1	X2	X3	X4
e1	light	boy	baby	red
e2	light	girl	baby	blond
e3	light	boy	child	chestnut
e4	medium	woman	teenager	chestnut
e5	medium	boy	child	red
e6	heavy	girl	child	blond
e7	heavy	man	teenager	red
e8	heavy	woman	teenager	chestnut
e9	heavy	man	adult	blond
e10	heavy	woman	adult	chestnut

It can also happen that no particular knowledge is available for a descriptor, like X4 which is a nominal descriptor. This knowledge is expressed by describing the set of values as an unstructured set: {blond, red, chestnut}. It may also happen that a value, such as "baby", is present in the tables, but that its relationships with the other values are not given.

5.2.2. Building R_i

Definition 1
Let us define a recognition function R as being described by:

$$R = (Xi = Vi) \ \& \ ... \ \& \ (Xk = Vk)$$

where the Xjs are descriptors and the Vjs are the values of these descriptors or disjunctions of possible values.

For example, the following are four different recognition functions:

X2 = man;
X2 = man ∨ woman ∨ boy;
X3 = [baby .. teenager];
(X2 = man) & (X3 = [baby .. teenager]).

Definition 2: Subset recognized by a recognition function.
This is the subset of examples such that the descriptors it contains have the same value as in the recognition function.

For instance, the recognition function R1: X2 = man ∨ woman ∨ boy recognizes the subset {e1, e3, e4, e5, e7, e8, e9, e10}. The function R2: (X1 = heavy ∨ light) & (X4 = red ∨ chestnut) recognizes {e1, e3, e7, e8, e10}.

Creation of a recognition function that recognizes e_i and rejects e_j: $G(e_i/e_j)$.
At the beginning, $G(e_i/e_j)$ is empty. Then, if the descriptor Xk has the same value in e_i and in e_j, then Xk does not occur in $G(e_i/e_j)$. If the descriptor Xk is different in e_i and e_j, let Vj be the value of Xk in e_j, then add the disjunct V_{ij} (Xk ≠ Vj) to $G(e_i/e_j)$.

Example: e1 and e2 differ by X2 and X4. In e2: X2 = girl and X4 = blond. It follows that G(e1/e2) = (X2 ≠ girl) ∨ (X4 ≠ blond). Similarly, one sees that G(e1/e4) = (X1 ≠ medium) ∨ (X2 ≠ woman) ∨ (X3 ≠ teenager) ∨ (X4 ≠ chestnut).

Comparison of an example and a set.
We are now able to compute the most general recognition function which recognizes the example e_i and rejects the set of examples $\{e_j\}$. Let us note it by $G(e_i/\{e_j\})$. This is what we have been calling R_i above.

For each e_k in $\{e_j\}$, calculate $G(e_i/e_k)$.
The conjunction of all the G obtained is $G(e_i/\{e_j\})$:

$R_i = G(e_i/\{e_j\}) = \&_{ek} \, G(e_i/e_k), \, e_k \in \{e_j\}$

Example: $G(e1/\{e2, e4\}) = G(e1/e2) \, \& \, G(e1/e4) = ((X2 \neq \text{girl}) \lor (X4 \neq \text{blond})) \, \& \, ((X1 \neq \text{medium}) \lor (X2 \neq \text{woman}) \lor (X3 \neq \text{teenager}) \lor (X4 \neq \text{chestnut}))$ which is then put in normal form.

5.2.3. Improving the Recognition Functions

These improvements are performed by using numerical information and background knowledge. Typical symbolic improvements minimize the number of disjuncts, or simplify the description. These improvements are always related to the application of a generalization rule [Michalski 1983].

Let us give two examples of such simplifications.

The "closing interval" rule states that several values of linear descriptors, such as lengths, may be unified by a single interval. For instance, suppose that Length(x, v) tells that the length of "x" is "v", and that the predicate Interv(x, min, max) says that min $\leq$ x $\leq$ max. One can replace the three conjuncts [Length(x, 5) & Length(x, 8) & Length(x, 10)] by the pair of conjuncts [Length(x, v) & Interv(v, 5, 10)]. This generalization obviously simplifies the recognition function.

The "detecting descriptor interdependence rule" states that one can try to replace sets of numerical values by a dependence they show. Suppose that the values a linear descriptor takes on in all descriptions are ordered in increasing order. Suppose further that the values of x and y increase together. Then, one can compute a new descriptor z = x/y, and check whether z is constant in all descriptions. If so, one will generalize the set of values of x and y by z = Const. This technique can be viewed as a discovery technique that detects new relationships between descriptors, but can also be viewed as an improvement of the recognition function, as we are doing here.

5.3. Structural Matching

5.3.1. Definition of Structural Matching (SM)

Two formulas structurally match if they are identical except for the constants and the variables that instantiate their predicates.

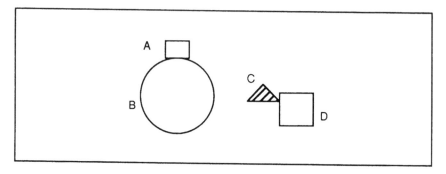

Fig. 3.14.   E1 is the left part of the figure. E2 is its right part.

More formally: Let E1 and E2 be two formulae, E1 structurally matches E2 iff there exists a formula C and two substitutions s_1, s_2 such that:

1-$s_1{}°C$ = E1 and $s_2{}°C$ = E2.
2-s_1 and s_2 never substitute a variable by a formula or a function.

where ° denotes the application of a substitution to a formula.
It must be understood that SM may be difficult and even undecidable. Nevertheless, in most cases, one can use the information coming from the other examples, in order to know how to orientate the proofs necessary to the application of this definition (for details, see [Vrain 1990]). Even if SM fails (which may often occur), the effects of the attempt to put into SM may be interesting. We say that two formulae have been SMatched when every possible property has been used in order to put them into SM. If the SM is a success, then SMatching is identical to putting into SM. Otherwise, SMatching keeps the best possible result in the direction of matching formulae.

5.3.2. A Simple Example of Successful SM

Let us consider the following two examples. E1 represents A piled on B, and E2 represents C and D, as shown in Fig. 3.14.
The examples can be described by the following formulas:

E1 = SQUARE(A) & SMALL(A) & UPON (A, B) & CIRCLE(B) & BIG(B) & UNDER(B, A)
E2 = TRIANGLE(C) & SMALL(C) & LEFT_SIDE(C, D) & SQUARE(D) & SMALL(D) & RIGHT_SIDE(D, C)

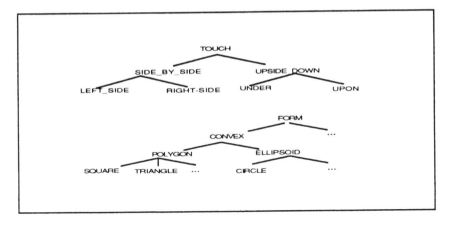

Fig. 3.15.   Taxonomies of generality relationships among the descriptors.

Let us suppose that the hierarchies shown in Fig. 3.15 are provided to the system. These taxonomies represent our semantic knowledge about the micro-world in which learning is taking place. The SM of E1 and E2 proceeds by transforming them into equivalent formulas E'1 and E'2, such that E'1 is equivalent to E1, and E'2 is equivalent to E2 in this micro-world (i.e., taking into account its semantics).
    When the process is completed, E'1 and E'2 are made of two parts:

One is a variablized version of E1 and E2, called the body of the SMatched formulae. When SM succeeds, the bodies of E'1 and E'2 are identical.
The other part, called the bindings (of the variables), gives all the conditions necessary for the body of each Ei' to be identical to the corresponding Ei.

The process starts as follows. Suppose that we know that the descriptors of the taxonomy "FORM" are "more interesting" than those of the taxonomy "TOUCH". We will therefore start attempting to match the "SQUARE" and "CIRCLE" of E1 with the "SQUARE" and "TRIANGLE" of E2. We replace "A" in E1 and "D" in E2 by the pseudo-variable "x", and record this value in the bindings. This gives (note that the list of bindings are between brackets):

E'1 = SQUARE(x) & SMALL(x) & UPON (x, B) & CIRCLE(B) & BIG(B) & UNDER(B,x) & [(x=A)]
E'2 = SQUARE(x) & SMALL(x) & RIGHT_SIDE(x, C) & TRIANGLE(C) & SMALL(C) & LEFT_SIDE(C, x) & [(x=D)]

The next step uses the taxonomies by matching the "CIRCLE" of E1 to the "TRIANGLE" of E2. One obtains:

E"1 = SQUARE(x) & SMALL(x) & UPON (x, y) & CONVEX(y) & BIG(y) & UNDER(y,x) & [(x=A) & (y=B)]
E"2 = SQUARE(x) & SMALL(x) & RIGHT_SIDE(x, y) & CONVEX(y) & SMALL(y) & LEFT_SIDE(y, x) & [(x=D) & (y=C)]

This example shows well that, once this SM step has been performed, the generalization step itself becomes trivial: we keep in the generalization all the bindings common to the SMatched formulas and drop all those not in common. In other words, this SM technique allows to reduce the well-known generalization rules [Michalski 1983, 1984] just to the "dropping condition rule" which becomes legal on SMatched formulas. All the induction power is in the dropping condition rule, all other rules are purely deductive. We must confess that formal proof of the above statement is still under research.

Since "BIG" and "SMALL" do not belong to a common taxonomy in our example, they will be dropped, and since "x" and "y" have always different values this will be noted in the final generalization which is:

Eg = SQUARE(x) & SMALL(x) & TOUCH(x, y) & CONVEX(y) & TOUCH(y,x) & [(x≠y]

In this example, one feels that one could use theorems like $\supset$x,y [TOUCH(x,y) $\Rightarrow$ TOUCH(y,x)] in order to improve the generalization.

5.3.3 - Using theorems to improve generalization
It can be easily guessed that using theorems can lead to many difficulties, since one enters the realm of Theorem Proving, which is well-known for being a good source of yet unsolved problems. In the case of SM, one is driven by the need to put the examples into a similar form, and the usual difficulties of Theorem Proving are somewhat eased. We cannot formally prove this point, but the following example, taken from [Vrain 1990] can at least illustrate our claim.

Starting from two examples that have no common predicates, we show that they nevertheless have a common generalization, found by using theorems that link the predicates.

Let the examples be:

E1 = MAMMALIAN(A) & BRED_ANIMAL(A)
E2 = TAME(B) & VIVIPAROUS(B)

to which the following theorems are joined:

R1: ⊃x [MAMMALIAN(x) & BRED_ANIMAL(x) ⇒ TAME(x)]
R2: ⊃x [TAME(x) & VIVIPAROUS(x) ⇒ MAMMALIAN(x)]
R3: ⊃x [TAME(x) ⇒ HARMLESS(x)]

where ⊃ is the universal quantification, & the logical conjunction, and ⇒ the logical implication.

The first step of SM is here trivial: we replace the constants by a variable x, and obtain the equivalent examples:

E'1 = MAMMALIAN(x) & BRED_ANIMAL(x) & [(x= A)]
E'2 = TAME(x) & VIVIPAROUS(x) [(x= B)]

Since the predicates have no common occurrence, we consider the first (this ordering is not significant, and just follows the one in which the examples are given) predicate of E'1: MAMMALIAN. We see that we can deduce this predicate from E2, using the rule R2. We get:

E"1 = MAMMALIAN*(x) & BRED_ANIMAL(x) & [(x= A)]
E"2 = TAME(x) & VIVIPAROUS(x) & MAMMALIAN**(x) & [(x= B)]

The MAMMALIAN of E'1 has been treated, this why it is marked by an * in E"1. The one of E"2 is derived from the use of theorems, this is why it is marked by **.

Again, using the order in which the examples are given, the next non-marked predicate is BRED_ANIMAL.

- No rule can be applied to E"2 to make explicit the presence of BRED_ANIMAL in it.
- Nevertheless, we remark that applying the rule R1 to E"1 uses the predicate concerned: BRED_ANIMAL. Checking the affect of this application, we see that it generates the atomic formula TAME(x) and that there is an occurrence of x in E"2 which matches this occurrence. Therefore, we conclude that we must apply R1 to E"1.

One obtains:

E"'-1 = MAMMALIAN*(x) & BRED_ANIMAL*(x) & TAME**(x) & [(x= A)]
E"'-2 = TAME*(x) & VIVIPAROUS(x) & MAMMALIAN**(x) & [(x= B)]

Now, the only unmatched predicate is VIVIPAROUS in E"2.

- No rules can be applied to E'''1 to make its presence explicit.
- The only rule which can be applied in E'''2, relative to VIVIPAROUS is R1. But, it would introduce the atomic formula MAMMALIAN(x), which is already matched since its instances are starred.

No other rule can be applied, we star the predicate VIVIPAROUS to remember that it has already been dealt with, obtaining:

E'''1 = MAMMALIAN*(x) & BRED_ANIMAL*(x) & TAME**(x) & [(x=A)]

E'''2 = TAME*(x) & VIVIPAROUS*(x) & MAMMALIAN**(x) & [(x= B)]

All possible occurrences have been dealt with, a complete SM is not possible, therefore the SMatching operation stops here.

Now, the generalization step is trivial: one drops the non-common occurrences, obtaining the generalization:

G = TAME(x) & MAMMALIAN(x)

This example shows well how potential infinite proof loops can be easily avoided, simply because they do not improve the SMatching state of the examples. More generally, one can use theorem proving techniques in order to improve the degree of similarity detected among the examples. Such a system is under development in our group [Vrain 1990]. It is not the concatenation of a classical theorem prover and of generalization algorithms, but instead is rigorously adapted to the kind of proofs required by Machine Learning. As an instance of its peculiarity (and of its incompleteness), it will not allow use of the same theorem twice during a given derivation.

6. THE QUESTIONNAIRE

From the preceding sections, the reader has a broad but fairly accurate description of what has been happening during these last few years in the field of ML (except the topic of automatic discovery which we did not include due to limited space). We are now ready to have a deeper look into the programs that are presently built, especially in Europe. We have been asking participants at European and French ML meetings (this explains why the French participation is so large) to answer the set of 17 questions below. The 16 first types of answers have been organized to fill up the tables of section 7, in the same order. The 17th one, i.e., the description of the algorithms fitted badly in a table and has therefore been placed after the tables.

The reader who is not familiar with any of the terms should refer to the glossary of terms given in the appendix. A summary of findings will be given in section 8. Section 8.2 provides a summary of the main features in the tables.

We give below the questionnaire, exactly as it was presented to the authors.

Answer NR when you think that the question is Not Relevant to your system. When you do not answer a question, NA (No Answer) will be added automatically.

6.1. Background knowledge
Knowledge intensive/knowledge poor. Representation of background knowledge

6.2. Meta-knowledge
Have you ways to express meta-knowledge? yes/no. If yes describe

6.3. Inputs
Representation of inputs (make precise their logic (0th, 1st order, ...), their semantic (what do their represent: Rules, conditions, ...?), the language you use (give an example if necessary)

6.4. Outputs
Representation of output (make precise their logic (0th, 1st order, ...), their semantic (what do their represent: Rules, conditions, ...?), the language you use (give an example if necessary)

6.5. Examples
You give positive only/negative only/positive + negative examples?
   Is the system able to generate itself new examples? yes/no

6.6. Incrementality
Describe how incrementality is achieved, if yes/no

6.7. Oracle
Do you use an oracle, a professor, or any other human interaction? yes/no. If yes, how the oracle gives information to the system?

6.8. Learning engine
Learning engine: induction/deduction/evaluation/merging of ...

6.9. What is learned?
What is given to, learned by the system?

6.10. Pre-clustering
Examples ordered prior to learning/system accept unordered examples. The system creates clusters of examples? yes/no.

6.11. Constructive
Is your system constructive in the meanings 1/2/3 that follow?
1. it uses its own learned knowledge to improve its performance,
2. it combines induction and deduction
3. it uses explanations to validate the newly acquired knowledge, and using it again.

6.12. Quality criterion
What are your own criteria to judge if your program has been behaving "well" or "bad"?

6.13. Stability
Is your program stable under learning twice the same data (in a different order, for instance)? yes/no
   If no describe.

6.14. Bias
Do you have an idea of the way you introduce preference bias? yes/no. If yes describe.

6.15. Theoretical model
Do you have a theoretical model relative to which you can validate your learning? yes/no. If yes describe.

6.16. Complexity
Can you evaluate the complexity of your algorithm as a function of the number n of examples, and their size s?

6.17. Algorithm
Give a short description (4 lines max) of the algorithm that does the learning (answer "NR" if you cannot do it in less than 4 lines). These descriptions will be given aside of the tables below.

7. ANSWERS

We received answers for 44 systems listed in table 3.3, in a non-significant ordering.

Table 3.

NAME OF SYSTEM	INSTITUTION	CONTACT PERSON(S)
ID3-TI	TURING INSTITUTE, Glasgow	T. Niblett
CN2	TURING INSTITUTE, Glasgow	P. Clark, T. Niblett
CIGOL	TURING INSTITUTE, Glasgow	S. Muggleton
ASSISTANT	Josef Stefan Inst., Ljubljana	B. Cestnik, I.Kononenko, I. Bratko
GINESYS	Josef Stefan Inst., Ljubljana	M. Gams
LogArt	Josef Stefan Inst., Ljubljana	B. Cestnik
NINA	Nixdorf, Paderborn	A. Ludwig, R Reppenhagen
DUCE	TURING INSTITUTE, Glasgow	S. Muggleton
ALEXIS	Telfonica Invest., Madrid	M. Nunez
SICLA	INRIA, Paris	E. Diday
MAKEY	Univ. Paris 6	J. Lebbe, R. Vignes
CONNECTIONISM (CONNEC)	(version of) EIHE, Paris	F. Fogelman
INSTIL	ESPRIT consortium, Europe	M. Manago
INDUCE	(version of) BRITISH AEROSPACE, Bristol	D. Hutber
ML-SMART	D. Informatica Univ. Torino	A. Giordana
ORLON	D. Informatica Univ. Torino	S. Bruno
RIGEL	D. Informatica Univ. Torino	R. Gemello
OGUST	LRI, Orsay	C. Vrain
DISCIPLE	LRI, Orsay & ITCI, Bucharest	G. Tecuci
BLIP	TUB, Berlin	K. Morik
EBG	MANY VERSIONS RE-IMPLEMENTED	

113

EBGF	LRI, Orsay	J. F. Puget
LIFE	LRI, Orsay	J. F. Puget
EBG (extended)	AI dept., Edinburgh	R. Desimone
ALEX	INSERM, Paris	B. Seroussi
MACHIN	LIFIA, Grenoble	C. de Sainte Marie
ALLY	IRISA, Rennes	J. Nicolas
CHARADE	LIFIA, Paris	J. G. Ganascia
CALM	CRIM, Montpellier	J. Quinqueton, J. Sallantin
PLAGE	CRIM, Montpellier	O. Gascuel
REFINER	DCS, Aberdeen	D. Sleeman, S. Sharma
INFER	DCS, Aberdeen	D. Sleeman, I. Ellery
MORE	Univ. Coimbra, Portugal	E. Costa
CLINT	DCS, Kat. Univ. Leuven	L. de Raedt
INDE	Univ. Amsterdam	P. Terpstra
CONCLAVE	AI-Lab VUB Brussels	W. Van de Welde
KBLA for Tonal Music	Res. Inst. for AI, Vienna	G. Widmer
AR4, BR3	Dept. Elec., King's Col. London	S. Kolabas
FM	DCS, City Univ. London	L. McCluskey
ZENO 1.1	Brainware Gmbit	J. Stender
MAVERICK	Brunel Univ.	A. MacDonald
BG	King's Col. London	A. Hutchinson
SEA	Ecole Polytechnique Palaiseau France	M. Sebag, M.Schoenauer
THUMBRULE	CADEPS, VUB Brussels	H. de Garis

114

	Background knowledge	Meta-knowledge	Inputs	Outputs	Examples
ID3-TI	know. poor	no meta-know	0th order classes of attrib.	0th order decision trees or rules	positive+negative don't generate
CN2	know. poor	no	0th order classes of attrib.	ordered list of 0th order IF-THEN rules	positive+negative don't generate
CIGOL	Know. intensive 1st order Horn clauses with function symbols	rules for know. transformation	rules in pure PROLOG	rules in pure PROLOG	positive+negative generate
ASSIS-TANT	know. poor	no	0th order, clas. & attr., can be continuous	0th order decision trees	positive+negative don't generate
GINESYS	know. poor	no	0th order classes and attributes	0th order rules	positive+negative don't generate
LogArt	know. poor	yes, rules about the particular domain	0th order classes and attributes	0th order rules with redundancy	positive+negative
NINA	know. poor , planned to become intensive	no	0th order classes and attributes	0th order rules extensions toward 1st order	positive+negative don't generate
DUCE	know. poor	no meta-know	0th order classes	decision trees or rules New concepts	positive+negative don't generate
ALEXIS	Taxonomies, costs, demons	no	attributes	decision tree	positive+negative
SICLA	know. poor	no meta-know	0th order, nume-rical or qualita-tive attr., classes	decision trees or rules clusters	p/n/p+n don't generate

MAKEY	know. poor, uses pre-order relations between descr. given by expert		0th order, conjunctions of multi-valued descriptors	0th order rules decision graphs	examples are concepts
CONNEC	explicit know. poor. Architecture of the net	no meta-know	0th order Boolean matrix	Boolean matrix don't generate	positive only
INSTIL	explicit know. intensive taxonomies, theorems, default values, belief coef.	object inheritance	1st order examples + classes	1st order decision trees, rules	p/n/p+n don't generate
INDUCE	generality taxonomies	no	0/1st order conj. of descr.	concepts	positive+negative don't generate
ML-SMART	knowledge intensive: frames, rules	yes, control rules	0th order, attr. of components of examples	graph of rules	positive+negative
ORLON	know. poor	yes, control rules	0th order Description of examples	taxonomy of concepts	positive only
RIGEL	know. poor	yes, control rules	1st order Description of examples	1st order rules	positive+negative
OGUST	explicit know. intensive taxonomies, theorems	no	1st order conj. of descr.	concepts	positive+negative don't generate
DISCIPLE	semantic net + rules, incomplete theory	no	rules 0th order	rules 1st order	yes, based on example gener.
BLIP	explicit know. intensive Horn clauses	meta & meta-meta rules	1st order facts	rules 1st order + concepts	NR
EBG	Complete theory supposed known	expl: operational. criterion, impl: proving strategy	1st order Example + validity proof	rules 1st order	positive only cant work with negative examples
EBGF	Complete theory given as Horn clauses with functional terms	expl: operational. criterion, impl: proving strategy	Example NOT implied by theory	failure condition	negative only

	Background knowledge	Meta-knowledge	Inputs	Outputs	Examples
LIFE	know. intensive STRIPS-like axioms	yes, determinism of some predicates	failures of plans	invariant patterns given as conjunctions	negative only examples generated by problem solving
EBG (ext.)	know. intensive rules of inference Existing proof tactics	proof plans	proofs in intuitionist type theory	proofs plans in 1st order sorted logic	positive only
ALEX	taxonomies of generality + descriptors complexity	implicit strategies	traces of solutions	heuristics for using operators	positive only generate examp.
MACHIN	knowledge poor. Can possibly use taxonomies of generality to be completed	certitude coef.	couples (situation, operator) certitude coef.	conditional rules + "certitudes"	positive + neg.
ALLY	Complete theory as clauses without functional terms	implicit: resolution strategies	1st order examples	1st order concepts	p/n/p+n
CHA-RADE	know. intensive. Horn clauses without func. Typed descriptors	Properties of the system of rules.	0th order examples as conjunctions	Approximate rules with certainty fact.	p/n/p+n not generate
CALM	Implicit. Descriptors properties	selection of formula by information	0th order examples as driven criteria	0th order, estimate of fit conjunctions	positive+negative concept/formula
REFINER	know. poor, some know. as semantic net	no	0th order classes attribute-value pairs	0th order decision criteria/ rules	positive+neg. don't generate
INFER	Know. poor	state transformation rules	0th order problem/ answers pairs	new (MAL)- rules	Not relevant
MORE	know. intensive Incomplete theory as Horn clauses	no	Horn clauses	Horn clauses	positive only
CLINT	know. intensive subsets of Horn clause logic	yes, given in the possible forms of the language	facts = 1st order ground instances	Horn clauses	positive+negative generates examples

117

	know. intensive subset of PROLOG clauses	type of errors type of arithmetic operator	set of PROLOG facts	PROLOG output	
INDE	know. intensive subset of PROLOG clauses	no			positive+negative don't generate
CONC-LAVE	contains hierarchies and causal model		problems + solutions	classifications + rules	positive+negative don't generate
KBLA Tonal Music	know. intensive: rules	heuristics for induction	examples of musical pieces	rules	positive+negative generate examples
AR4, BR3	know. intensive Descriptive + procedural as PROLOG clauses facts and rules	pseudo-2nd order	examples of particle reactions	particle reactions	positive only generate ex.
FM	strips-like operators	control rules	initial & goal states	solutions + control rules	NR. Uses planning trace as +/- exam.
ZENO 1.1	know. poor	no	0th order classes of attributes	0th order rules	positive+negative generates ex.
MAVE-RICK	know. poor	no	independent lang. arbitrary	representation predictions in input language	stream of positive ex. only don't generate
BG	know. poor	no	context free grammar, sequence of characters	context free grammar	positive only don't generate
SEA	know. poor	no	set of (attribute, discrete values)-class	0th order rules	positive+negative don't generate
THUMB-RULE	know. poor	no	conjuncts of attr/val pairs	disjuncts of conjuncts of attr/val pairs	positive+negative don't generate

	Incrementality	Oracle	Learning engine	What is learned?	Pre-clustering	Constructive?
ID3-TI	no	no	Entropy optim. in each class, particularization, post-pruning	given: classes -> decision tree for the classes	no	no
CN2	no	no	general-to-specific BEAM research	lists of rules for the classes	no	no
CIGOL	yes, incremental generalization of new example	yes tells if stat. are ok gives name	induction & evaluation	given: description of situations learned: relations between objects	yes	yes, 1,2,3
ASSIS-TANT	no	yes, force attribute selection	induction	given examples -> decision tree for the classes	yes	yes, 1
GINESYS	no	no	induction	given: classes -> rules for the classes	yes	yes, 1
LogArt	yes, by rejecting no longer valid rules	no	induction	given examples -> selected rules: 2/3 attributes, credibility	yes	yes, 1
NINA	no	no	Entropy optim. in each class, particularization	rules	no	no
DUCE	NA	yes	generalization	given classes -> concepts	yes	
ALEXIS	no	no	trade off cost/entropy optimization	given classes -> decision tree for the classes	yes	
SICLA	depends of chosen algorithms	no	distance minimization in R^n	un-clustered examples -> clustered	no	no
MAKEY	no	no	Deduction Evaluation	concepts descriptions -> concept identif. methods	no	no

System	incrementality				
CONNEC	yes. Based on	backpropagation of gradient, coefficient optimization	Connections weights giving desired output for given input	yes	
INSTIL	no	Entropy optim. in each class + generalization	decision tree for the classes + concepts at node of dec. tree	yes	
INDUCE	no	aggregation of samples algorithm	rules discriminating pos/neg examples	yes	
ML-SMART	no	merging induction and deduction	concept discriminant descriptions	yes	yes, 2
ORLON	no	induction	new concepts	no	NA
RIGEL	no	induction	concept discriminant descriptions	no	NA
OGUST	possible yes	generalization	given: clustered examples, get: intentional descriptions	yes	
DISCI PLE	yes, based on oracle	EBL + SBL + analogy + retrieving from semantic net	problem solving rules from example solutions	no	
BLIP	yes	deduction + induction from facts	new rules from facts new concepts from contradictions	no create clust.	yes, 1 + 2
EBG	yes	theorem proving	more efficient rule	no	
EBGF	yes	merging of deduction & partial evaluation, some induction	recognition of dead ends described as generalizations of counter-examples	NR	yes, 2 3 is NR

	Incrementality	Oracle	Learning engine	What is learned?	Pre-clustering	Constructive?
LIFE	yes, after each backtrack of the pb solver	no	deduction	invariant patterns	yes	yes 1 3 is NR
EBG (extd)	NR	no	precondition analysis	in: example proofs out: proof plans	NR	yes, 1&3 (potentially)
ALEX	yes	no	recognition of sequences of operators, Generalization	in: solutions, out: heuristics for using operators	no	NA
MACHIN	yes, based on incrementality	no	over-generalization followed by refinement evaluation of belief coef.	learns heuristics for using operators + builds net of operator dependencies	no	yes, 1 & 3
ALLY	yes	no	theorem proving	given: clustered examples, get: intentional descriptions	yes	
CHARADE	no	no	top-down induction	obtains rules valid on the whole set of examples	no	NA
CALM	one step of incrementality	no	induction, evaluation statistics	gets rules in favor of, and against a concept	no	yes, 2
REFINER	yes	yes, correc of errors	induction	learns distinguishing features of concepts	no	yes, 1 & 3
INFER	no	yes	induction	new (MAL)-rules	yes	yes, 1
MORE	yes	yes, completes theory	resolution	set of beliefs describing the cognitive model of an agent	yes	NA
CLINT	yes, based on	classifies generated examples	induction + deduction	concept descriptions	no	yes, 1 & 2

System						
INDE	yes, incremental version of AQ	yes	merging of EBL, AQ15, error propagation	refined rules, new rules	no	NA
CONC-LAVE	yes	yes, user's observations	induction, evaluation of classification rules	classifications + rules	yes	yes, 1 & 2
KBLA for Tonal Music	yes, new rules added to KB	yes teacher	deduction + induction + analogy	rules for completing counter-point pieces	yes	yes, 1 & 2 & 3
AR4, BR3	yes, belief revision	of little use	induction + deduction	find valid particle reactions, composion of elementary particles, quantum principles	no	yes, 1 & 2 & 3
FM	yes	yes supplies tasks	induction + deduction	control rules	NR	yes, 1 & 2
ZENO 1.1	no	no	induction	rules for classes	no	no
MAVE-RICK	yes, based on	no	SBL + generalization + statistics	sequence of event description -> tree of predictive rules -> sequence of predictions	no	no, 1 is planned
BG	yes	no	induction + statistical clustering	new descriptors	no	yes, 1
SEA	no	no	induction	rules	no	NR (=no)
THUMB-RULE	no	no	induction	class rules	no	no

	Quality criterion	Stability	Bias	Theoretical model	Complexity
ID3-TI	comparison with classical ES, and human perf., Tree simplicity	yes, checked by partitioning exa.	yes, user-supplied pruning parameter	Information theory, statistics	n, s
CN2	comparison of rules accuracy & simpl with other systems	yes	user selects desired statistical significance of rules	statistics	p=nb attrib/ex. $O(n, p)$
CIGOL	performance against pre-computed solut. Human evaluation	yes	yes, statistical tests based on complexity in time ans space	Resolution theorem proving, and information theory	can be computed
ASSIS-TANT	check on testing set	yes	no	Information theory	$O(n*s)$
GINESYS	check on testing set comparison with statistical methods	yes	yes, "impurity", "error-estimate", parameters, etc ...	Information theory, Bayes' formula, statistics	$O(n)$, $O(s^x)$, x>1
LogArt	check on testing set	yes	no	Bayes formula without attribute independence assumption	$O(n*s^2)$
NINA	comparison with classical ES and human performance	yes	not explicit	Information theory	$O(n*s)$
DUCE	NA (see CIGOL)	NA	NA	NA	NA
ALEXIS	comparison with classical ES and hum. performance	yes	NA	Information theory	NA
SICLA	yes, evaluated by missing axes of inertia	yes	choice in measure of distance	Statistics	$n \leq$ complex. $\leq n^4$
MAKEY	form of the graphs	yes	attribute pre-order quality criterion	no	evaluation: $n^2 * \log n * s$

CONNEC	check on examples	yes	values of coefficients before learning + net architecture	connectionism	NA
INSTIL	expert + user tree depth, branching factor	Costs, reliability, explanations pref. by expert	beliefs on importance & reliability of descriptors	Information theory + object oriented languages	linear in n, s^2
INDUCE	expert	yes	simplicity measures choice of seed	no	NA
ML-SMART	performance on new examples, fit in theory and compactness	yes	explicit control rules	no	not checked
ORLON	increasing coherence inside clusters	yes	explicit control rules	categorization theory	not checked
RIGEL	completeness and consistency	yes	explicit control rules	no	not checked
OGUST	obtain most particular generalization	not checked	ordering of descriptors by importance	structural matching	not checked
DISCIPLE	expert	not stable	explicit in knowl. base	no	not checked
BLIP	better structuration of domain theory	not stable, depends on meeting neg. example	meta-knowledge	theorem proving	linear with nb of facts
EBG	fit in theory	should	choice of strategy of proofs	theorem proving	same as theo. proving
EBGF	simplicity	yes	choice of strategy of proofs, operationality criterion	negation as failure	s=nb of proofs* size of proofs linear in s
LIFE	large changes in problem solving efficiency	yes	no	mathematical induction applied to constructive domains	s=nb of proofs* size of proofs s*log(s)

	Quality criterion	Stability	Bias	Theoretical model	Complexity
EBG (extended)	empirical success in guiding future	NR proofs	choice of primitive tactics determine	theorem proving possible proof plans	not checked
ALEX	analysis of success or failure on new	not checked	measure of resembl-ance between two expressions	no	not checked
MACHIN	large decrease in error rate = nb of backtrck/ nb of steps in solution	no	choice of initially given legal operators	no, under research	little information
ALLY	fit in theory	yes	tries to avoid biasing	theorem proving	NA
CHA-RADE	compactness of result Expert's advice	yes	properties of the looked for set of rules	structural matching boolean lattices	n, but dep. most on s (nb descrip)
CALM	Expert's advice and appreciation	yes, in average	explicit: adjustable parameters	Inference in majority logics	$O(s*n^3)$
REFINER	expert's advice	yes	favors frequently obs. features and conjuncts	no	does not depends on n ans s
INFER	expert's advice	yes	favors simpler combinations	no	large but unknown
MORE	performance with new examples	NR	partition of the theory	no	not checked
CLINT	user's advice	no, if 2 definit. are consistent	choosing the language	identification limit proven	not checked
INDE	performance on test + correction of theory	yes	by introducing over-general rules, rather than over-specific	no	not checked

125

System			type of abstraction		
CON-CLAVE	practical correctness simplicity	yes		yes: pb solving + definition of learning goal & quality	not checked
KBLA for Tonal Music	improve interaction less stupid questions. Compare to theory	yes	expl: heuristics impl: language + model of pb solving	theory of counterpoint	not yet checked
AR4, BR3	compare with human performance	yes	no	no	NA
FM	performance improv.	no, qualitity of control rules dep. of tasks	simple tasks must be presented first	EBL	not checked
ZENO 1.1	performance on test data and expert's advice	yes	simplicity	depends on algorithms	not checked
MAVE-RICK	performance on test data, rate of convergence, info. gain	yes	concepts given in knowledge base	Information theory Automata theory Markov theory	possible to compute not done
BG	simplicity	yes	parameters in algo. statistical confidence levels	hill climbing in space of grammars	not checked
SEA	performance on test data	yes, only minor changes due to ordering	order of input and attributes	no	$O(s*n^3)$
THUMB-RULE	simplicity	not sure	form of inputs and outputs	no	$nc=$nb obj. p. class $N_r=$av. nb. partial rules per class $O(2^{nc}N_r)$

Algorithm descriptions

ID3-TI. 1. Grow decision tree (traditional ID3 algorithm)
        2. REPEAT Prune tree leaves
        UNTIL Pruning termination criteria met

CN2. Procedure CN2. REPEAT: find a good rule UNTIL no more can be found.
Procedure FIND-A-RULE:
1. Start with the general rule "everything is <class>"
2. Perform general-to-specific beam search for better specializations
3. Return the most accurate, statistically significant rule found

CIGOL. REPEAT UNTIL terminated by user
        READ EXAMPLE
        GENERALIZE/ADD NEW PREDICATES
        TEST AGAINST ORACLE

ASSISTANT. Modified ID3 with the following additional mechanisms. Forward pruning and post-pruning to fight noise; binarisation of attributes; selection of good learning instances (optional).

GINESYS. Similar to CN2, AQ15. Basic modifications are that it constructs redundant rules, combines with Bayes rules approximation, enables parameter specifications.

LogArt. Generate all the rules with up to 2/3 attributes. Eject bad rules. Sort the remaining rules according to their credibility.

NINA. ID3 + continuous attributes, "don't care" values, optional binarisation, attribute/attribute comparisons, post-pruning.

DUCE. See Cigol.

ALEXIS. The method contains two algorithms FM and GS. FM searches through a frame-based structure of the examples and send a selected frame in form of a table to GS. GS builds a decision tree, according to an economy criterion. FM joins the decision trees and builds a macro-tree.

SICLA. Clustering looks for a set of concepts the most linked with the original descriptors. Statistical criteria are used, they are based on inertia or chi-square criteria. The process is iterative and starts from initial stars. Discriminant methods generate step by step a decision tree. They work top-to-bottom and choose the most discriminant descriptor at each step.

MAKEY. Creation of decision graphs top-to-bottom. At each step the most discriminant descriptor is chosen, guided by background and meta-knowledge. The consequences of each node built are evaluated.

CONNEC. NR.

INSTIL. ID3 algorithm written in an object oriented language so as to use background knowledge during the building of the decision tree and being able to evaluate the consequence of chosing a particular node.

INDUCE. Based on Michalski's INDUCE, with extras
1. Handling of noisy data.
2. Dynamic type taxonomies created
3. Different search heuristics used.

ML-SMART. Top-down search guided by declarative heuristics and domain theory.

ORLON. Top-down clustering.

RIGEL. Specialization and generalization steps interleaved.

OGUST. First order generalization algorithm using background knowledge to increase the degree of matching of the examples. Goal oriented theorem proving is used to discover relevant properties.

KBG. First order generalization algorithm using background knowledge to increase the degree of matching of the examples. Forward chaining, i.e. exhaustive propagation of properties, is used to discover properties that increase the degree of matching.

DISCIPLE. The system receives examples of problem solving from its user. It over-generalizes the examples and applies this over-generalization to the data base in order to find instances of the over-generalization. These instances are proposed to the user as tentative rules. When the user validate them, we say that they are positive examples, when s/he rejects them, we speak of a negative one. This set of positive and negative examples is used by a classical generalizer to produce a more refined generalization.

APT. Same as DISCIPLE.

BLIP. Learns by hypothesis generation, then hypothesis testing; a hypothesis is a rule-schema instantiated by predicates. Test patterns attached to each rule-schema (characteristic situations) are used to find positive and negative examples in the facts, including derived facts.

A threshold and ratio is used to decide wether to reject or to accept the hypothesis as a new rule.

EBG. Prove relevance of training example to concept. Prune proof trace by "operationality criterion". Generalize pruned proof trace.

EBGF.
1. Try each possible proof of the goal using the negative example
2. For each proof, collect all the leaves (as in EBG)
3. Reduce the number of literals
4. Remove negations by using determinism (inductive step)

LIFE. Construction and analysis of a dependency graph describing the effects of an action on the given situation.

EBG (extended). Match object level steps against meta-level tactics. Use back-propagation to fit meta-level tactics together into proof plan.

ALEX. NR.

MACHIN. As long as there is a problem to solve or that knowledge needs evaluation

Use knowledge to choose a solving step

Evaluate result using knowledge (subjective) and, if possible, feedback on previous experience (objective).

IF subjective evaluation is negative, or impossible, or different from objective evaluation

THEN create rule (that will need evaluation)

ALLY. NA.

CHARADE. Example and description spaces are viewed as a boolean lattice. Regularities are extracted using correspondence between these two spaces. A set of rules is built by exploration of the description space, computing regularities for each description and using rule system properties (i.e meta-knowledge) in order to constrain the exploration.

CALM. Three mechanisms build logical formulas. Expansion is the generation of candidate formulas. Selection is selection of good formulas. Compression is clustering equivalent formulas.

PLAGE. Close to branch-and-bound algorithms. Breadth first and top-down search of version space.

REFINER. Identifies distinguishing features of each concept (in differential diagnosis), and generalizes them.

INFER. Infers mal-rules to cover gaps in input student traces.

MORE. NR.

CLINT. Receive positive examples from user.

WHILE possible DO

Ask for classification of examples generated by system

IF an example is said to be positive THEN generalize

INDE. NA.

CONCLAVE. Maximally generalize on purpose, specialize when forced to.

KBLA (Tonal Music). Integration of deductive and inductive learning under the paradigm of "learning as compilation of explanations". The system builds explanations of situations and generalizes the explanation structure. Explanation steps can be deductive (as in EBG), by analogy (vs. determinations) or weaker forms of similarity. This provides a flexible integration of deductive and inductive effects.

AR4, BR3. NR

FM. From weakest precondition WP=WP(O1, O2, ..., ON, S, G) of the sequence {O1, ..., On} where G is the goal. Generalize WP. Form $WP^° \approx$

{WP & On's precondition}. Specialize WP° until it discriminates against alternative instantiations of On in original proof trace.

ZENO 1.1. Based on AQ, ACLS, CN2. Significant modifications follow. in ACLS there is a different handling of numerical attributes. In CN2 thereare different strategies for handling noise.

MAVERICK. Sequences of descriptions at appropriate level of generality collected in a tree. Selection of predictions based on accuracy/evidential trade-off.

BG. REPEAT

SCAN TEXT (repeatedly PARSE)

ANALYSE PARSES OF TEXT

AUGMENT GRAMMAR, on basis of statistical properties of parses plus other properties.

SEA. For each example, all discriminant generalization of maximal generality are investigated. Rules are selected according to an exception threshold to optimize performance on a test-set.

THUMBRULE. Similar to Diday's "CABRO", optimized to generate class rules (disjuncts of conjuncts of attr/val pairs) with minimum length (i.e. total number of pairs).

8.   A BRIEF DISCUSSION OF THE TABLES

We intend to prepare an extensive discussion of these tables with the help authors of the systems. Since this work is not yet completed, we shall present here the main features only in the answers.

8.1. Background Knowledge

The distribution of knowledge intensive vs poor is roughly even, which reflects the strong influence of the ID3-like class of algorithms in Europe. It is generally given under the form of a set of clauses, and/or taxonomies.

8.2. Meta-knowledge

As expected, the same count shows with meta-knowledge, most of the authors building knowledge intensive systems are conscious of the meta-knowledge they introduce, implicitly or explicitly. There is a large variety of ways to represent this meta-knowledge.

8.3. Inputs

The inputs of a ML program may take many different forms, like attributes pairs, example descriptions etc, and can represent many different kinds of knowledge, like rules, conditions, etc. Most authors, except the EBL community, work with 0th order inputs. One should understand that the answers "Horn clauses" or "PROLOG facts" mean 1st order, while "attribute" or "pairs (attribute, values)" mean 0th order.

8.4. Outputs

Again, most authors, except in the EBL community, work with 0th order inputs.

A very striking fact is that very few programs (explicitly: DISCIPLE and CLINT, perhaps also CONCLAVE, KBLA, AR4, and FM) generalize 0th order inputs into 1st order outputs. The reason for that feature lies in the fact that going from 0th order representations (e.g., while describing aircraft flights, flight_altitude = 500 feet) to instantiated 1st order representations (e.g., flight(altitude, 500)) is not at all trivial. One has to choose a predicate (e.g., is it flight(x, y) as in our example, or altitude(u, v) as it could well have been? What makes the choice of 500 (z, t) "obviously" wrong?). In more complicated cases, one has also to choose appropriate skolem functions. These choices made, generalizing instantiated first order expressions to variablized expressions is easier, if not trivial.

8.5. Examples

There is a smooth distribution of systems that use positive only, or both positive and negative examples. In the tables, "generate" means that the concerned system generates examples, and "don't generate" means that they do not.

About 10% only of the systems generate their own examples.

8.6. Incrementality

Incrementality is an important issue since no real life learning system can expect to get the information it needs all at a time. Nevertheless, almost one half of the systems do not bear incrementality. This means that most of the existing systems are learning tools expected to be included in a ML system, rather than ML systems by themselves.

SICLA is a special case since it is the generic name of many numeric data analysis programs, some of them incremental, others not, hence the puzzling answer "depends on the chosen algorithm".

8.7. Oracle

An oracle, as defined in the glossary, is a human giving reliable instructions or information to the system.

Systems which do not use oracles can be classified as pure ML, while systems which do have some kind of KA component could be used as a KA system. As we shall see in section 9, there are few systems originating from the KA community that make use of ML. The tables show that more than 20% of the ML programs show this important KA feature. This explain why some of them are listed in both tables.

8.8. Learning engine

Only 10 systems rely on pure induction, only 3 of them rely on pure deduction. It is quite striking to see that most actual ML programs rely on combinations of induction or deduction with some other technique.

8.9. What is learned?

Obviously, the outputs are learned from the inputs. Therefore, this question may be often redundant with the 3rd and the 4th. This is wrong since a learning program can perfectly well have the same kind of inputs and outputs, and nevertheless perform learning by improving the inputs in some sense.

8.10. Pre-clustering

This column must be well understood. A "yes" means that "yes, the system needs pre-clustering", i.e., that it is not able to perform clustering unaided. A "no" means "no, there is no need for pre-clustering", i.e., the system can perform it alone.

8.11. Constructive

This question was meant to test the intuitive definition of constructive (definition 1), i.e., that the system uses its own learned knowledge to improve its performance, versus the most elaborate last two definitions. It turns out that among the 19 authors that claim having constructive programs, only 6 claim using definition 1 in isolation. It means that, apart from belonging to any school of thought, definition 1 is not the only possible definition of constructive learning. This gives some experimental evidence to Michalski's classification of constructive learning in Figure 1.

8.12. Quality criterion

All the programs that evaluate their performance by checking on a testing set, or by human evaluation are strongly oriented towards induction. Conversely, almost all the programs able to generate their own internal system of reference for quality contain a deductive component, or have a strong numeric flavor, like SICLA and MAKEY.

8.13. Stability (eg, order-dependence)

One could be amazed at the relatively high proportion of programs that are not stable. This feature is usually considered as unbearable in software engineering. In the context of learning, however, one must be aware that it might well be quite realistic, and yet unstable, in other words, that lack of stability is not so negative as in the usual software environment.

8.14. Bias
This question is somewhat redundant with question 2 about meta-knowledge since bias is implicit meta-knowledge. Most authors liked this redundancy since they acknowledged bias even when they did not acknowledge meta-knowledge.

8.15. Theoretical model
As for question 13, one could be amazed at the relatively high proportion of programs without theoretical model, which is unthinkable in most of computer science. As expected, inductive programs that do not use information theory or statistics do not often have a theoretical model. Formal models of induction, as the one presented in ALLY, are still seldom available.

8.1.6. Complexity
We did not expect so many authors to have evaluated the complexity of their algorithms, since ML is essentially NP-complete. Some authors refused to evaluate the complexity of their algorithm as a function of the number n of examples, and of their size s. They asked for other definitions of the complexity, as reflected in the tables.

When looking at this column, one should remember that "NA" means that the author did answer the question, while "not checked" means that the author said himself that he did not check it.

8.17. Algorithm
The only comment we shall give here is an apology for having some descriptions in such a bad shape. Hopefully, the future interaction we are planning with the authors will increase the understandability of their descriptions.

9. DESCRIPTION OF SOME LEARNING KNOWLEDGE ACQUISITION (KA) SYSTEMS

We end this paper with a short description of current work in KA. Firstly, one should recall that KA is knowledge acquisition from an expert that tries to transfer directly, explicitly, his knowledge to some automatic reasoning system, an expert system in most of the cases. In contrast, ML is an indirect, implicit, transfer of knowledge from examples of the expert's behavior into such an automatic system. Therefore, KA concentrates on friendly interfaces between humans and systems (and reduces its learning to "learning by being told"), while ML concentrates on deep analysis of the expert's behavior (and reduces its interface friendliness to acquiring knowledge by observation, without questioning). It is quite obvious that the ultimate interface will some day be KAML since asking clever questions to an expert requires much learning abilities, and ML could use a bit more of learning by being told. In the current state-of-the-art, however, the two fields are quite disjoint, and there are very few KA systems explicitly referring to ML possibilities. During EKAW'88 [Boose, Gaines & Linster 1988], we tried to describe how systems behave that do include ML. It shows the relatively high importance given in Europe to coupling KA and ML since most of the US works depend on friendly interfaces rather than on ML.

9.1. Model changes
KA uses a model of the problem in order to ask the user questions. This model can be improved automatically or, alternatively, through interactions with the user. If there is any ML at all, there should be some model changes during the KA phase. This piece of information specifies the kind of changes made by each system.

9.2. Reliability
In KA systems, the reliability of the system becomes even more stringent than in other ML systems since the user is supposed to control the acquisition phase.

9.3. Domain
This piece of information specifies the domain of application of the KA system.

9.4. Method
What is the ML method used by the system?

9.5. Cooperation
This describes how the user and the system interact when they meet a problem. For instance, an inductive system can perform itself easy induction steps and rely on its user when it fails.

9.6. Source
What is the source of the knowledge, how do you "spy" your user?

9.7. Authors
"Esfahani": L. Esfahani, F.N. Teskey, Brighton Polytechnic,

Table 4.

	Model Changes	Reliability	Domain	Method	Cooperation	Source
Esfahani	Intermediate concepts Consistency checking + relevant restructuration	User's advice	Engineering Planning Design	Rule checking	Exp. gives rules, gets rules	Form filling Graphical notes from experts
Hoppe	New rules	User's advice	UNIX file handling	constants ->variables Recog. task patterns	monitoring the user	Dialog protocols
Someren	Deep rules → shallow ones user's + Model refined & completed	yes + solving advice	Problem EBL + SBL	trivial stat. error propagation	Exp. gives examples, gets rules + Deep model imprv.	Sets of examples
Moller	Add knowledge	Knowldge engineer advice	Independent of	Matching context	K. eng. selects proposals, gets descriptions	Texts domain models
Boy	Problem reformulation	User's adv + expert+ questions	Intelligent Tutoring Systems	Chunking	Based on coop. System must guess user's intentions	Tutoring sessions
Sebag	No model	measured by system	Failure detection Maintenance	SBL	no cooperation system gets examples, gives rules	Incomplete arrays of parameters
Tecuci	Completion of user's model	User's advice	Problem solving	Integration EBL+SBL + analogy	Based on coop. 0th order rules -> 1st order ones	Solutions of problems + questions to user
Witten	New programs	Unknown	Graphics	generalization	Based on coop. Strong metaphor of teaching	Execution traces

"Hoppe": H. U. Hoppe, GMD IPSI, Darmstadt,
"Someren": P.P. Terpstra, M. W. van Someren, Univ. Amsterdam,
"Moller": J.U. Moller, Univ. Hamburg,
"Boy": G.A. Boy, N. Nuss, ONERA Toulouse,
"Sebag": M. Sebag, M. Schoenauer, Ecole Polytechnique Paris,
"Tecuci": Y. Kodratoff & G. Tecuci, LRI Orsay, George Mason Univ. &
ITCI Bucharest
"Witten": D.L. Maulsby, I.H. Witten Univ. Calgary.

10. CONCLUSION

10.1. General comments on the tables, and comparison with the US state-of-the-art

Obviously, many of these programs are inspired by approaches stemming from statistics, numeric data analysis, or information compression. Such are the numerous improvements to ID3 and AQ. The improvements which European researchers seem to be very keen on are concerned mainly with the understandability of the results, and the use of background knowledge. This is in accordance with part of the US efforts (for example Michalski's school), but also at right angles with the more technical improvements looked for in the US. For instance, windowing or incremental use of ID3 are quite popular in the US, while they draw little attention in Europe. Conversely, we know of no US knowledge based ID3, while INSTIL and NINA are two European approaches to this problem. Similarly, there are no US equivalent to the large efforts devoted by the Turing Institute and the Ljubljiana school to the understandability of the results.

Another typical feature of the European school is its tendency to develop new generalization algorithms, like ALLY, BLIP, CHARADE, CLINT, CONCLAVE, OGUST while the US school seems to be more satisfied with the algorithms it has been developing.

When coming to multi-strategy learning, the US school seems to concentrate on merging Explanation-Based Learning and empirical learning, while the European school is more versatile as shown by ALEX, using analogy and generalization, BLIP using induction in a deductive environment, and DISCIPLE merging EBL, empirical learning and analogy.

Another important difference between the US and the European approach lies in the relatively low influence in Europe of the "EBL revolution". Except the pioneering works of the Edinburgh school led by

Bundy, EBL caught quite late, and little work has been done to improve EBL except by LRI's EBGF and Edinburgh's EBG.

Finally, we have the feeling that the classifier approach, together with the genetic algorithms which have been mainly developed in the US, have been scarcely catching in Europe. MACHIN and CALM can be seen (even though their authors may strongly contest our opinion) as representative of the classifier approach in that sense that they rely very much on performance evaluation in order to learn. Their essential difference and originality lies in the amount of work done to fit with the domain representation, instead of complying to the classical strings of bit representations.

10.2.   General Conclusion

One can see the field as divided in two broad areas.

The first one stems from a statistical approach, and relies on information compression techniques. Its most prominent representative are Ross Quinlan, Donald Michie, and Ivan Bratko. However, instead of working only on refinements of the statistical methods, it insists on the necessity to optimize the understandability of the results as well. For instance, a decision tree of less statistical significance may be preferred on the grounds that the tree itself is more readable. Since this approach is easy to implement, close to existing statistical methods, it received much attention from the industry, and it was used with success in a number of applications. However, one must notice that understandability is quite a social notion. For instance, Ivan Bratko (personal communication) remarks that in some cases the experts claim that rules are "easier to understand", while others with different backgrounds will claim the same of decision trees.

The second one stems from the symbolic, AI oriented approach. Its founders are the editors of the books "Machine Learning: An AI approach, volumes 1 and 2", Ryszard Michalski, Tom Mitchell, and Jaime Carbonell. This approach divides itself into two pieces, depending on the inference method they use, deductive or inductive. Even though it comes from less well established grounds, since it is newer, the AI approach has received much more academic interest (especially in the U.S.A.) than the first one. One should notice however that few of its applications can claim to be of industrial interest.

These two areas are far from having exhausted their research field. Nevertheless, it seems that merging the inductive and the deductive approach, called constructive learning by Michalski (see, for instance [Michalski & Ko 1988], will constitute a third topic of the AI approach.

We have been involved in the ESPRIT project INSTIL, which merges the information compression technique together with the AI inductive techniques, and we are participating in the Machine Learning Toolbox which gathers into a single toolbox ten of the programs described in this paper. We have been also involved in the ESPRIT project ALPES, in which there is a significant emphasis on program synthesis from specifications. This allowed us to start a study on predicate synthesis from their specification, in order to build the recursive expression of a totally new predicate, as long as one is able to define it. This explains why we would like to conclude this paper by stressing the fact that merging constructive learning in the sense of Michalski, and systems using inductive and deductive inference, information compression, and also, predicate synthesis techniques, will be the next step implementing programs that learn efficiently in real life problems.

ACKNOWLEDGEMENTS

A first version of this questionnaire has been set up by J. G. Ganascia and N. Helft [Ganascia and Helft 1988]. An other source of information was the first meeting of the Machine Learning Toolbox ESPRIT2 project. In that meeting were present representatives of British Aerospace, CGE Marcoussis, DCS of Aberdeen Univ., INRIA Rocquencourt, INTELLISOFT, LRI of Univ. Paris Sud, Nixdorf, Univ. Paris Dauphine. The knowledge acquisition questionnaire was elaborated during the 2nd European Knowledge Acquisition Workshop in Bonn, with the help of Ian Witten, and the participants to the workshop. Many questionnaires have been filled up during EWSL-88 held at the Turing Institute. All our discussions with Ryszard Michalski concerning the structure of ML were also very helpful. A careful proof reading by K. Dontas helped to improve notably this paper. The editor's remarks were also quite helpful.

This research was sponsored by grants from French "PRC-GRECO IA", CEC COST-13 and ESPRIT programmes. The research was partly done in the Artificial Center of George Mason University. Research activities of this center are supported in part by the Defence Advanced Research Projects Agency under grant, administered by the Office of Naval Research, No. N00014-87-K-0874, in part by the Office of Naval Research under grant No. N00014-88-K-0226, and in part by the Office of Naval Research under grant No. N00014-88-K-0397.

REFERENCES

[Arbab & Michie 1985] Arbab, B., Michie, D. "Generating rules from examples". *Proc IJCAI–85* , A. Joshi (Ed.), Morgan Kaufmann 1985, Los Altos, pp.631-633.

[Barr and Feigenbaum 1981] Barr A., Feigenbaum E. A. *The Handbook of Artificial Intelligence, Volume 1*, William Kaufmann Inc., Los Altos, 1981.

[Bareiss, Porter & Weir 1990] Bareiss E. R., Porter B. W., Wier C. C. "Protos: An Exemplar-Based Learning Apprentice", to appear in *Machine Learning: An Artificial Intelligence Approach, Volume 3*, Y. Kodratoff, R.S. Michalski (Eds.), Morgan Kaufmann 1990, pp.112-139.

[Boose, Gaines & Linster 1988] Boose J., Gaines B., Linster M. (Eds.) Proceedings of EKAW'88, Bonn June 1988, GMD-Studien 143.

[Bratko & Lavrac 1987] Bratko, I., Lavrac, N. *Progress in Machine Learning*, Sigma Press, Wilslow 1987.

[Bratko, 1988] Bratko, I. *Unpublished set of lectures at European summer school on Machine Learning*, Les Arcs, 1988, France.

[Carbonell 1983] Carbonell J.G., "Learning by Analogy: Formulating and Generalizing Plans from Past Experience" in R.S. Michalski, J. G. Carbonell, T. M. Mitchell (Eds.), *Machine Learning: An Artificial Intelligence Approach*, Morgan Kaufmann 1983, pp. 137-159.

[Carbonell 1986] Carbonell J.G., "Derivational Analogy: A Theory of Reconstructive Problem Solving and Expertise Acquisition", in R.S. Michalski, J. G. Carbonell, T. M. Mitchell (Eds.), *Machine Learning: An Artificial Intelligence Approach, Volume II*, Morgan Kaufmann 1986, pp. 371-392.

[Chouraqui 1985] Chouraqui E., Construction of a model for reasoning by analogy. *Progress in Artificial Intelligence*, Steels L., Campbell J.A. (Ed.), Ellis Horwood, London, p.169-183, 1985.

[Corlett 1983] Corlett, R. "Explaining induced decision trees", *Research and development in expert systems*, M.A. Bramer (Ed.), Cambridge University Press, 1983.

[Cox & Pietrzykowski 1986] Cox P.T., Pietrzykowski T., Causes for Events: Their Computation and Applications. *Proceedings of the eighth International Conference on Automated Deduction*. Oxford 1986.

[Davies 1987] Davies T. R., Russell S. T., A logical Approach to Reasoning by Analogy. *Proceedings of IJCAI 87*, p 264-269, Milan. Los Altos, Ca: Morgan Kaufmann.

[DeJong & Mooney 1986] DeJong G., Mooney R., Explanation-Based Learning: An alternative View. *Machine Learning ,1*, p.145-176. Kluwer Academic Publishers.

[Duval & Kodratoff 1989] Duval B., Kodratoff Y. "A Tool for the Management of Incomplete Theories: Reasoning about explanations" to appear in *Machine Learning, Meta-Reasoning, Logic*, P. Brazdil eds, Pitman 1989.

[Ganascia and Helft 1988] Ganascia J.-G., Helft N., "Evaluation des Systèmes d'Apprentissage", *Actes Journées Franìaises sur l'Apprentissage*, E. Chouraqui editor, Cassis May 1988, CNRS Marseilles 1988, pp. 3-20.

[Genesereth and Nilsson 1985] Genesereth, M. R., Nilsson N. J., *Logical foundations of Artificial Intelligence*, Morgan Kaufmann 1985.

[Kedar-Cabelli & McCarty 1987] Kedar-Cabelli, S.T., McCarty, L.T. "Explanation-based generalization as resolution theorm proving", *Proceedings of the Fourth International Machine Learning Workshop* 1987, pp.383-389.

[Kodratoff et al. 1984] Kodratoff, Y., Ganascia, J.G., Clavieras, B., Bollinger T. and Tecuci, G., "Careful Generalization for Concept Learning", *Proceedings of the Sixth European Conference on Artificial Intelligence*, Pisa, 1984, pp.483-504.

Also in Advances in Artificial Intelligence, T. O'Shea (Ed.) North-Holland Amsterdam (1985), pp. 229-238.

[Kodratoff & Ganascia 1986] Kodratoff, Y., Ganascia, J.G., "Improving the generalization step in learning" In: *Machine Learning: An Artificial Intelligence Approach, Volume II*, R.S. Michalski, J.G. Carbonell, & T.M. Mitchell (Eds.), Morgan Kaufmann 1986, Los Altos, pp.215-244.

[Kodratoff 1988] Kodratoff Y. *Introduction to Machine Learning*, Pitman 1988.

[Kodratoff & Michalski 1990] Kodratoff Y., Michalski R. S. (Eds). *Machine Learning: An Artificial Intelligence Approach, Volume III*, Morgan Kaufmann 1989.

[Kodratoff & Tecuci 1987] Kodratoff, Y., Tecuci, G., "Techniques of Design and DISCIPLE Learning Apprentice". *International Journal of Expert Systems 1*, no. 1, 1987, pp. 39-66.

[Kodratoff 1990a] Kodratoff, Y., "Combining similarity and causality in creative analogy" *Proc. ECAI-90*, L. Carlucci Aiello (Ed.), Pitman, London, 1990, pp.398-403.

[Kodratoff 1990b] Kodratoff, Y., "Using abductive recovery of failed proofs for problem solving by analogy", In: *Machine Learning, Proc. 7th International Conference on ML*, B.W. Porter, & R.J. Mooney (Eds.), Morgan Kaufmann, Palo Alto, Calif., 1990, pp.295-303.

[Laird, Newell, & Rosenbloom 1986] Laird, J.E., Newell, A., Rosenbloom, P.L., *Universal subgoaling and chunking: the automatic generation and learning of goal hierarchies*, Kluwer, Dordrecht, 1986.

[Laird, Newell, & Rosenbloom 1987] Laird, J.E., Rosenbloom, P.L., Newell, A., "Soar: An architecture for general intelligence", *AI Journal 33*, 1987, pp. 1-64.

[Michalski 1984] Michalski R., "Inductive learning as rule-guided transformation of symbolic descriptions A theory and implementation", in Automatic Program Construction Techniques, Biermann, Guiho & Kodratoff editors, Macmillan Publishing Company 1984, pp. 517-552.

[Michalski et al. 1983] R.S. Michalski, J. G. Carbonell, T. M. Mitchell (Eds.), *Machine Learning: An Artificial Intelligence Approach*, Morgan Kaufmann 1983.

[Michalski et al. 1986] R.S. Michalski, J. G. Carbonell, T. M. Mitchell (Eds.), *Machine Learning: An Artificial Intelligence Approach, Volume II*, Morgan Kaufmann 1986.

[Michalski 1983] Michalski, R.S., "A Theory and Methodology of Inductive Learning," in *Machine Learning: An Artificial Intelligence Approach, Vol. 1*, R.S. Michalski, J.G. Carbonell, T.M. Mitchell (eds.), Tioga, Palo Alto, CA, 1983, pp. 83-134.

[Michalski & Ko 1988] Michalski, R. S., Ko H., "On the Nature of Explanation", *Proceedings of the Symposium on the Explanation-based Learning*, Stanford University, March 21-23, 1988.

[Michalski & Stepp 1983] Michalski R.S., Stepp R.E, "Learning from observation: Conceptual Clustering" in *Machine Learning: An Artificial Intelligence Approach*, R.S. Michalski, J.G. Carbonell, T.M. Mitchell (Eds.), Morgan Kaufmann 1983, pp 331-363.

[Mitchell 1982] Mitchell T. "Generalization as Search", *Artificial Intelligence 18*, 1982, 203-226.

[Mitchell et al. 1986] Mitchell T., Keller R., Kedar-Cabelli S. Explanation-Based Generalization: A Unifying View. *Machine Learning Journal ,1*, p. 47-80., Kluwer Academic Publishers.

[Nicolas 1988] Nicolas J., "Consistency and Preference Criteria for Generalization Languages handling Negation and Disjunction", *Proc. ECAI-88*, Y. Kodratoff eds, Pitman 1988, pp. 402-407.

[Nunez 1988] Nunez, M., "Economic induction / A case study", *Proc. 3rd EWSL 1988*, Glasgow Oct. 1988, D. Sleeman (Ed.), Pitman, London, pp. 139-145.

[Puget 1987] Puget, J.-F. Apprentissage de plans à partir de preuves. *Proceedings of AFCET 1987*, Antibes.

[Priedetis 1988] Priedetis, A., (Ed.) *Analogica*, Pitman, London, 1988.

[Quinlan 1983] Quinlan J.R., "Learning Efficient Classification Procedures and their Application to Chess End Games" in *Machine Learning: An Artificial Intelligence Approach*, R.S. Michalski, J.G. Carbonell, T.M. Mitchell (Eds.), Morgan Kaufmann 1983, pp 463-482

[Schank 1986] Schank R. C. (1986) Unpublished set of conferences made during the winter 1986, at University Paris 7.

[Schank 1987] Shank R. C. *Explanation Patterns: Understanding mechanically and Creatively*, Ablex Publishing Company, (1987).

[Schank and Abelson 1977] Schank R. C., Abelson R. P., *Scripts, Plans, Goals, and Understanding*, Lawrence Erlbaum, Hillsdale, N.J. (1977).

[Schank & Kass 1990] Schank, R.C., Kass, A. "Explanations, machine learning, and creativity", In: *Machine learning: An artificial intelligence approach, Volume 3*, Y. Kodratoff & R.s. Michalski (eds.), Morgan Kaufmann, 1990, pp. 31-48.

[Siqueira & Puget 1988] Siqueira, J., Puget, J.-F., "Explanation-based generalization of failures", *Proc. ECAI-1988*, Y. Kodratoff (Ed.), Pitman, London.

[Vrain 1990] Vrain, C., "OGUST: A system that learns using domain properties expressed as theorems", In: *Machine learning, an artificial approach, Volume III*, Y. Kodratoff & R. Michalski (Eds.), Morgan Kaufmann, 1990, pp.360-382.

[Waldinger 1977] Waldinger R. "Achieving several goals simultaneously", In: *Machine Intelligence 8*, E. Elcock & D. Michie (Eds.), Ellis Horwood, London, 1977.

[Winston 1982] Winston, P.H., "Learning new principles from precedents and exercises", *AI Journal 19*, 1982, pp. 325-350.

APPENDIX: GLOSSARY

Abduction. Process of completing a failed reasoning. See more details in section 3.

Analogy. A way of reasoning or learning by comparing similarities and differences between two examples, situations, or solutions. Should be based on a base (A, B) and a target (A', B'). A and A' are similar, B depends causally on A. The analogy process attempts to see how much the similarity between A and A' entails either similarity between B and B', or causal dependence of B' from A'.

Axiomatic/scheme-based learning. In both cases, the learning makes use of background knowledge. They differ by the degree of rigor in this knowledge. Axiomatic learning takes place from very rigorous (i.e. coherent and complete) knowledge, while scheme-based learning can use somewhat incoherent knowledge, and certainly does use incomplete knowledge.

Background knowledge. Knowledge about the field in which the learning is taking place. The difference of knowledge intensive vs knowledge poor refers to the amount of available explicit background knowledge. There exists almost no real learning system that does not use specific knowledge about the domain in which the learning is proceeding. Nevertheless, most systems encode this knowledge in the representation used, and/or in the way the computations are performed. In these cases, we say that the knowledge is implicit, that is these systems are knowledge poor.

Bias. Biasing amounts to make choices that are not easily tracked. As a consequence, as in the case of background knowledge, implicit biasing is a current practice in ML. The first example of explicit biasing happened as choices of nodes in a taxonomy of generality.

Bottom-up/top-down generalization. Bottom-up generalization performs "real" generalizations, (see this word) while top-down generalization actually performs a particularization. See more details in section 1.3.

Constructive learning. R. S. Michalski first defined constructive learning as combining induction and deduction (meaning 2 in the tables). We propose here two other definitions in order to test whether they were entailed by each other in practice since, from an intuitive point of view, "constructive" may also mean that the system uses its own learned knowledge to improve its performance. We suggest a third possible definition based on the iterative use of explanations to validate the newly acquired knowledge.

Cooperation. Describes the cooperation between a system and its human user or teacher. For instance, an inductive system can perform the easy induction steps and rely on its user when it fails.

Domain. Describes the domain of application of the system.

Explanation-Based Generalization (EBG): given complete knowledge of the domain in which learning takes place, and one example of correct behavior, EBG will prove the example correctness, analyze the trace of this proof, and generalize it just enough to respect its sufficiency. The analysis is done in terms of a so-called criterion of operationality describing the operators on which one has to focus attention. Notice that EBG does not perform second order generalization, i.e. it does not generalize the operators themselves but only the expressions to which they apply.

Explanation-Based Learning (EBL): This approach to ML analyses one positive instance and keeps its significant features. It uses formal knowledge to prove the instance validity as in Explanation-Based Generalization but also uses informal knowledge, given as schemes of a semantic net, in order to determine the level of generality at which the operators themselves should be used.

Explanation-Based Generalization by Failure (EBGF): an EBG in which a complete theory of the possible failures has been included. When a failure is detected, the theory about it is used to recover from it.

Empirical Learning from Examples: often called "Similarity-Based Learning" because its goal is the detection of similarities between sets of positive examples (in order to find their complete Description) and between sets of negative examples (the negation of similarities between negative examples constitute Coherent Descriptions of the positive examples). Keep in mind that inductive hypotheses thus drawn should be as reasonable as possible. This reasonableness follows from background knowledge added to the examples. It follows that, contrary to widespread belief, the difference between Analytical Learning and Empirical Learning does not stem from the quantity of background knowledge they use but from the way they use it. Analytical Learning uses it for the detection of intra-example properties while Empirical Learning uses it for the detection of inter-example properties.

Empirical Learning from Observation: similar to Empirical Learning from Examples except that the examples are provided by the system itself. It may happen that a strong example generator may then compensate for a weaker system of Empirical Learning. For instance, it is often the case that background knowledge is left implicit in the examples generator, and the empirical learning proceeds without it.

Empirical learning. Inductive learning performed without using background knowledge. With all its imperfections, empirical learning can be very useful as a quick first "glance" at possible relationships.

Examples. Systems often learn from examples. When these examples illustrate a behavior described as correct by the oracle, they are called positive examples. When they illustrate a behavior described as incorrect by the oracle, they are called negative examples. When the system is able to generate new examples itself, they can be either positive and negative. Experimentation and/or expert interrogation will classify them.

Generalization: Extending the scope of a concept description to include more instances (the opposite of specialization). This term is sometimes used as a noun, synonymous with concept description. Provides an intentional description of a concept, as opposed to conceptual clustering that provides an extensional description of a concept.

Incrementality. Induction systems often learn from a given set of examples. Adding a new example requires starting over again from scratch. Incrementality is achieved when the system is able to evolve smoothly when new examples are added.

Inductive learning of decision trees. Known as "ID3" from its first implementation in the ML sub-field. Builds a decision tree on criteria that optimize both the entropy and some understandability of the tree itself.

Knowledge acquisition (KA). Body of techniques intending to help an expert to express his/her knowledge in a form usable by a system expert. It goes from a simple window management to elaborated interrogation forms. Most KA systems do not use machine learning at all since their goal is collecting information from a human expert, rather than asking the machine itself to acquire knowledge. In other words, one can say the KA systems are intended to perform clever rote learning. Nevertheless, it is clear that if some learning is performed by the system itself then it can avoid asking tedious questions to the user, and/or it can ask real clever questions. Therefore, ML is becoming a strong component of KA, without being merged with it.

Knowledge intensive induction. Induction making explicit use of background knowledge, see more explanations in section 3.

Knowledge intensive / knowledge poor. See Background knowledge.

Learning engine. Inductive learning makes use of induction as its main learning engine, while deductive learning (and explanation)-based learning also) make use of deduction. In many cases, evaluation of the program performances is a learning engine as well.

Learning from examples / observations. Since observations can always be seen as kinds of examples, the only real difference between these two kinds of learning lie in the way examples are presented. If they are already clustered, and the concept to learn is already described (i.e. one learns a new description) then one speaks of learning from examples. One speaks of learning from observations in all other cases. In the last case, the system must be able to create its own clusters of examples.

Meta-knowledge. Meta-knowledge contains rules that state properties of the knowledge itself. For instance, if knowledge is made of rules, then meta-rules state relations among the rules, or strategies for using the rules.

Model changes The KA uses a model of the problem in order to ask questions to its user. This model can be improved automatically or, alternatively, through interactions with the user.

Oracle. An oracle is a human interacting with the system by answering selected questions asked by the system.

Pre-clustering. Examples can be ordered prior to learning, or, alternately, the system may accept unordered examples. In the tables, a "yes" means that the relevant system needs pre-clustering, while a "no" means that the system is able to accept unordered examples.

Quality criterion. They are the criteria to judge if a program has been behaving "well" or "badly". Such criteria, or some of them, can be internal to the system which will decide itself the quality of its results.

Stability. It is not always the case that a program gives identical output twice on the same data, in a different order, for instance. A learning program may well be very reliable even though it is not perfectly stable.

NOTE

1. A rigorous description of the different modes of induction is presently undertaken by the author. The following presentation is so much simplified that it might puzzle some theory-oriented readers. They are kindly required to allow their intuition to play with the examples, instead of letting them stopped by the lack of precision.

CHAPTER 4

A Perspective
on Machine Vision

Hans-Hellmut Nagel
Fakultät für Informatik der Universität Karlsruhe (TH)
and
Fraunhofer-Institut für Informations- und
Datenverarbeitung (IITB), Fraunhoferstr. 1,
D-7500 Karlsruhe 1/Federal Republic of Germany

ABSTRACT

A framework for machine vision is sketched in order to discuss current research problems in the evaluation of digitized images. The recognition and description of objects, of their shape, texture, and other properties, as well as a description of their static spatial relations with respect to each other and to the recording camera, all require complex system-internal representations.

This framework is extended to discuss the investigation of image sequences which permit temporal developments within the field of view of a recording camera to be captured. The temporal variation of image intensities can be exploited to infer temporal variations of geometrical relations in space. Although this raises a number of very interesting research problems, this contribution emphasises a different research avenue: linking the results to be extracted from an image sequence to descriptions at the level of natural language sentences. In the field of computational linguistics, considerable experience has been accumulated in the construction of internal representations for natural language statements about the world. Tapping this experience for research into the evaluation of image sequences is expected to be of advantage.

1. INTRODUCTION

Vision is a multi-faceted notion, usually related to a capability of living creatures. It enables them to sense the electromagnetic radiation in or near the visible spectrum in order to evaluate the state of their environment and its changes relative to their own position. Given the importance of such a capability for the survival of mobile creatures, it is not surprising that at least one third of the human brain is directly related to vision. Moreover, vision and behavior are intimately intertwined. It has been estimated, for example, that over 90% of the information required to drive a car is mediated to a driver by vision (Avant et al., 1986).

Behavior is understood to consist of much more than merely reflex-like (re-)actions; it comprises sequences of more or less complex activities each of which may consist of multiple elementary actions. Although complex activities cannot, in general, be executed in an instant, they are nevertheless likely to be specified subconsciously or consciously as an entity and to be represented as a, potentially generic, unit. This consideration suggests that behavior is represented within a living creature at multiple levels of abstraction in order to facilitate the management of representations for complex activities. Such abstractions are needed, too, for communication among humans. It is, therefore, obvious that we should expect close relationships between the abstractions developed for the internal representation of behavior and concepts expressed by natural language.

Given a close relationship between vision and behavior, vision thus also influences the concepts by which we describe states as well as behavior—both our own and the ones of our environment. In other words, vision deeply shapes the way we think and thus perceive ourselves.

In view of these considerations, it appears appropriate to distinguish between the notion of vision as related to living creatures and a new term—machine vision. This latter notion is introduced specifically in order to avoid unconsidered and thus, at least as yet, potentially unwarranted associations between biological vision and preliminary results of algorithmic investigations. Machine vision is understood to denote algorithmic approaches, which relate spatio-temporal samples of radiation obtained by a two-dimensional transducer, for example a video-camera, to a system-internal representation of the depicted section of the environment, in order to derive plans and parameters for actions by an engineered, as opposed to a grown, system. This contribution will be restricted to machine vision. References to the literature about vision in the more general sense can be found, for example, in Marr (1982), Braddick and Sleigh (1983), Feldman (1985), and Levine (1985). Additional references can be found in the section on "Biological Motion Processing" of the 1989 Workshop on Motion (Anonymous, 1989).

2. A PERSPECTIVE ON MACHINE VISION

Research in machine vision is influenced by many disciplines, some of which are shown explicitly in Fig. 4.1. Applications oriented research in machine vision has been included here in order to draw attention to the fact that machine vision cannot avoid being influenced by the continuous, substantial decreases in the cost/performance ratio of processors, memories, and peripherals. Such developments can lead to substantial shifts in emphasis among different approaches, for example the fact that today the tendency prevails to compute a convolution directly even for sizeable masks rather than indirectly via a multiplication of Fourier-transforms using FFT.

As a result, solutions to application problems are increasingly derived by brute force: the fact, however, that an approach works in a particular instance—often under not fully explored circumstances—may obscure rather than provide insights which could be generalized.

Considering the six human sensory modalities, there are not yet substantial research activities to investigate artificial intelligence equivalents to the senses for temperature, taste, and smell—i.e. as far as the development of complex, system-internal representations and their exploitation is concerned. Touch has recently become an active field of research, especially in conjunction with robotics and machine vision in so-called multi-sensor approaches. Artificial touch is, however, restricted to a local, relatively coarse-resolution exploration of the system environment—similar to its biological analogue.

It is now interesting to note that the two sensory modalities which provide the largest spatio-temporal bandwidth for the exploration of the environment—namely seeing and hearing—are so far treated quite differently by research:

Neurophysiology	Psychophysics	Psychology
Computer Science	**Machine Vision**	Artificial intelligence
Pattern Recognition	Control Theory	Applications research

Fig. 4.1.

- machine vision is more or less expected to cope with the totality of real-world complexity, whereas
- speech recognition so far concentrates on temporarily localized sections, namely isolated spoken words or at most short spoken sentences.

In systems for the analysis of spoken natural language texts, the internal representations of the discourse world thus tend to be of limited complexity. System approaches for the automatic understanding of written natural language texts, having no problems with a signal-to-symbol transition, investigate much more complex descriptions. As a consequence, such systems require internal representations at a more abstract, conceptual level.

Longer image sequences present a particular challenge for machine vision to cope with complex spatio-temporal situations in the visually perceptible world. It thus appears plausible to assume that the internal representations of machine vision systems, for extended image sequences, are more closely related to representations investigated in natural language text understanding, than to those investigated in speech recognition. This hypothesis will be taken up again in a later section which discusses the extension of machine vision for non-stationary scenes.

In view of the relation between vision and behavior discussed in Section 1, machine vision should be seen as an algorithmic approach to close a feedback loop between imaging sensors and effectors manipulating the system environment in order to achieve certain goal states. This point of view, by the way, is the justification to include control theory among the disciplines in Fig. 4.1. Fig. 4.2a is an attempt to sketch important ingredients of situations in which biological vision is normally applied whereas Fig. 4.2b illustrates the severely restricted circumstances under which most current machine vision approaches are pursued:

- a single camera,
- fixed to a stationary platform,
- no possibilities for the vision system to manipulate the scene, components or the illumination in order to improve the conditions for detection or disambiguation.

Up until recently these circumstances have been essentially dictated by the difficulties to record digitally, store, manipulate and evaluate more than a few image frames at costs which could be afforded by a university research laboratory. Moreover, it still is a challenge to develop

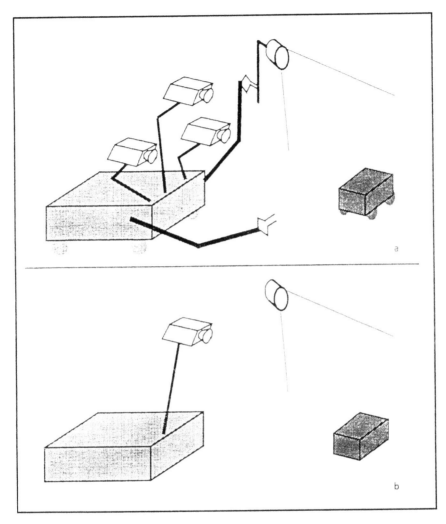

Fig. 4.2.   a) Important ingredients of how machine vision should be applied in analogy to biological vision: several adjustable cameras (eyes in head), mounted on a mobile platform (body) equipped with manipulators (hands) which can influence the illumination of a scene with mobile objects: b) How current machine vision approaches are deprived compared to biological vision: a single camera, fixed on a stationary platform, with no manipulators and no influence on the illumination of a scene which contains only stationary objects.

Name	Purpose	Primitives
Image(s)	Represents intensity	Intensity value at each point in the image
Primal Sketch	Makes explicit important information about the two-dimensional image, primarily the intensity changes there and their geometrical distribution and organization	Zero-crossings, blobs, terminations and discontinuities, edge segments, virtual lines, groups, curvilinear organization, boundaries
$2\frac{1}{2}$ - D Sketch	Makes explicit the orientation and rough depth of the visible surfaces, and contours of discontinuities in these quantities in a viewer-centered coordinate system	Local surface orientation (the "needles" primitives), distance from viewer, discontinuities in depth, discontinuities in surface orientation
3 - D model representation	Describes shapes and their spatial organization in an object-centered coordinate frame, using a modular hierarchical representation that includes volumetric primitives (i.e. primitives that represent the volume of space that a shape occupies) as well as surface primitives	3 -D models arranged hierarchically, each one based on a spatial configuration of a few sticks or axes, to which volumetric or surface shape primitives are attached

Fig. 4.3. Representational framework for deriving shape information from images according to Marr (1982, Table 1-1).

a machine vision approach which has an, at least in some restricted sense generic, ability to recognize shapes from first principles, in distinction to pictorial pattern recognition approaches which rely on humans to define (application dependent) "features" in order to recognize objects by "classification". It is one of the challenges for machine vision to provide a theory of feature extraction from images.

This state of affairs is reflected by the fact that Marr (1982) essentially concentrates on the derivation and representation of shape descriptions—see Fig. 4.3. Temporal variations of gray value distributions are discussed by Marr primarily from the point of view what their evaluation could contribute within this shape-oriented

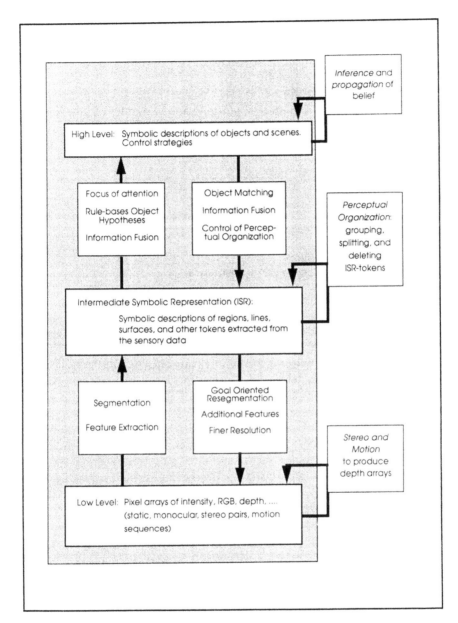

Fig. 4.4.   Processing and Control across multiple levels of representation in the VISIONS system according to Hanson and Riseman (1988).

representational framework. Even in quite recent descriptions of long-standing system approaches towards machine vision such as the VISIONS system (Hanson and Riseman, 1988), this point of view still prevails, as exemplified by Fig. 4.4.

Many additional aspects of current research in machine vision are treated in a special issue on Computer Vision edited by Li and Kender (1988). The following section discusses principal aspects of current machine vision approaches such as those indicated by Figs. 4.3 and 4.4 in terms which facilitate subsequent conceptual extensions in order to cope with set-ups where the deprivations illustrated by Fig. 4.2 no longer dominate.

3.   BASICS OF MACHINE VISION FOR STATIONARY SCENES

Kanade (1978) suggested discussing machine vision as a process iterating between six major process-states—see Fig. 4.5. The generic description mentioned in Fig. 4.5 represents a parameterized model for the scene, complemented by parameterized models for the scene illumination and the imaging system by Nagel (1979, 1987). The initial step of top-down or model-based approach to machine vision according to Kanade (1978) has to generate a-priori hypotheses about a scene depicted by an image to be analyzed. This step generates—at least partially—instantiated models or prototypes by restricting parameters to a subrange or a fixed value within the range of variability provided in the generic description. This instantiation subprocess comprises, too, the fixing of parameters defining the properties of the imaging system, including its position and attitude relative to the scene.

The next step employs this knowledge to generate a scene projection in the form of a synthetic image which can be compared subsequently to the image of the depicted scene obtained by the transducer. Significant differences between the observed and the synthetic image are converted to picture domain cues which are combined with knowledge about the physics of image generation in order to derive cues in the three-dimensional scene domain. The scene domain cues are in turn exploited to select more appropriate models or parameter subranges from the generic description for the scene under investigation in order to improve the prototypes to be used during the next iteration around this loop. This iterative search for appropriate model instantiations can be terminated as soon as no significant differences remain between the scene projection and the original digitized image.

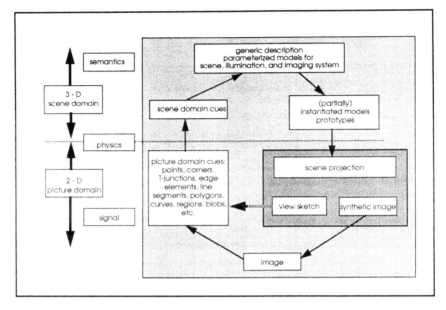

Fig. 4.5.   Machine vision according to Kanade (1978), with modifications by Nagel (1979).

Instead of generating a synthetic image, one might generate some abstraction thereof, the view sketch, which represents the depicted scene only by certain features such as

- points (associated with gray value extrema, corners, or line intersections),
- edge elements,
- straight or curved line segments,
- blobs,
- regions, etc.

These are to be compared directly to picture domain cues or features extracted from the original image.

The extraction of features may be considered, too, as the initial step in a bottom-up or data-driven approach to machine vision if not enough a-priori knowledge about the depicted scene is available to generate at least partially appropriate prototypes right from the beginning. The discussion about "seeds of perception" by Brady (1987) may be considered as an example for such an approach. Only under severely restricted conditions, however, is it possible to jump directly from simple picture domain cues to correct conclusions about an object in the scene:

"verification vision" investigated, e.g., by Bolles (1977) in connection with the fast inspection of parts or assemblies may serve as an example.

Space restrictions do not permit a full review of approaches detailing the steps indicated in Fig. 4.5. Numerous approaches have been discussed for the evaluation of gray value transitions ("edge elements"). Canny (1986) formulated and evaluated performance criteria with the goal to optimize simultaneously the detection and the localization of gray value transitions. A recursive variation thereof has been investigated by Deriche (1987). The, essentially one-dimensional, shape of the gray value transition to be looked for is rarely specified explicitly although in most cases an implicit specification corresponds to a smoothed step transition. Nalwa and Binford (1986; see, too, Nalwa 1987) provide an example for an explicitly assumed transition function, in their case a tanh. Such one-dimensional approaches have difficulties with strongly bent gray value transition fronts, for example around corners. Apart from a lack of a model for the two-dimensional aspects of gray value transition fronts, the degree of smoothing implied by current convolution-type edge-detectors is another cause for concern. Fleck (1987) and more recently (1988) investigate alternative approaches which are based on a more direct evaluation of pixel aggregates in order to overcome these difficulties.

The aggregation of edge elements into contour segments usually proceeds via search of approximating straight line segments, for example see Ayache and Lustman (1987). Although the choice of other segment shapes such as circular, elliptic or parabolic arcs have been investigated, too, the greater number of parameters to be estimated for such segment shapes and an increased susceptibility to noise has not yet resulted in robust approaches which are widely accepted. In addition to one-dimensional aggregates, two-dimensional aggregations attempt to derive region descriptions on the basis of either gray value or texture distributions which are assumed to be characterizable by few parameters only—ideally by just one, namely a constant. If the scene is suitably chosen, such essentially heuristic approaches may work fairly well, but they are rather brittle against variations which are unavoidable in less controlled scenes.

The transition from 2-D picture domain cues to 3-D scene domain cues can be mediated by a considerable variety of both geometric as well as radiometric cues. A good example for assumptions which facilitate such a transition from 2-D to 3-D is provided by the interpretation of line drawings of purely polyhedral scenes. Culminating a quarter of a century of research into this specialized problem area, Sugihara (1986) presents a theory for the interpretation of such line drawings provided the objects in the scene comply with additional constraints such that every edge in the scene is shared by exactly two faces.

Unless it is known a priori that only a special set of objects may appear in the scene, an attempt is made first to extract surface patches, based on some hypothesis about local surface smoothness in the scene. Among the geometry oriented approaches, shape from relative point and line segment positions or—more generally—from contour or texture variations (see, e.g., Kanatani and Chou 1989) as well as shape from stereo (see, e.g., Faugeras 1989) have been widely investigated. Among the radiometric approaches, shape from shading has been studied intensively, although most have been based on simplifying assumptions about the reflectance properties of surfaces in the depicted scene—see, e.g. Horn (1986), Horn and Brooks (1986, 1989). Recently, gray value distributions associated with highlights from surfaces with specular reflectance have been investigated in order to obtain shape cues (Healey and Binford, 1987). Very encouraging results have recently been obtained by an approach to infer shape attributes from systematic color variations (Klinker et al., 1988). Depth estimates based on calibrated focus readings—see, e.g., Krotkov (1987)—for local image areas may serve to corroborate shape hypotheses obtained by other approaches.

In general, however, the current state for estimating aggregated picture domain cues as well as scene domain cues could be characterized as being analogous to that for edge element extraction: computationally feasible algorithms have known deficiencies with respect to modelling significant structural variations in the sensed data, but the inclusion of admittedly necessary extensions quickly leads to algorithms which require computational resources not yet generally available. Moreover, many of the approaches still have a heuristic flavor in that they concentrate overly on single aspects. Although the integration of cues from different sources—see, for example, Aloimonos and Shulman (1989)—has long been recognized as important, it has not turned out to be an easy solution.

4. EXTENSION OF MACHINE VISION FOR NON-STATIONARY SCENES

The image interpretation loop of Fig. 4.5 lends itself to an extension which allows one to incorporate the representation and evaluation of temporal developments in an intuitively plausible manner—see Fig. 4.6. The most simple approach would consist in just reproducing the loop from Fig. 4.5 for each frame time t, t+1, t+2, ... This would correspond to activating this interpretation loop at each frame time without regard to the information acquired at the preceding time instant. It is much

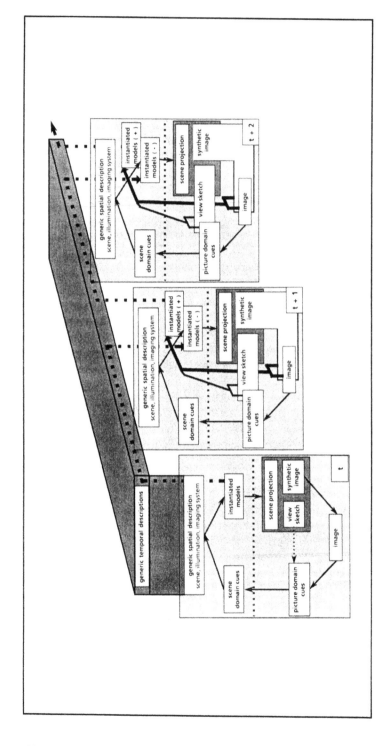

Fig. 4.6 Extension of the image interpretation loop from Fig. 4.5 to incorporate temporal developments. Generic temporal descriptions are used to predict the instantiated 3-D models from time t to time t+1. These *expected* 3-D models, denoted by (-), are then used to predict either a synthetic image or a view sketch at time t+1 which are subsequently compared to the observed image or the picture domain cues, respectively, at time t+1 in order to update the representation of the instantiated models at time t+1, denoted by (+).

more advantageous to employ some generic description of admissable temporal developments in order to predict the 3-D instantiated models expected for time t+1 from those instantiated at time t. These 3-D model instances—denoted in Fig. 4.6 by (-)—are then used to derive predicted scene projections for frame time t+1, based on the predicted position and attitude of the recording camera as well as of the illumination. The scene projection could be either in the form of a synthetic image or of a view sketch. The resulting scene projection is subsequently compared to the image recorded at frame time t+1 or to the picture domain cues derived from it in order to estimate corrections for the parameters in the 3-D instantiated models at frame time t+1. These updated 3-D models are designated in Fig. 4.6 by appending a (+).

It should be kept in mind that the image interpretation loops of Fig. 4.6 actually comprise two intra-frame loops:

1 The loop introduced in Fig. 4.5 which evaluates the static geometric and radiometric properties at each frame time. This includes the instantiation of object models which become visible for the first time in that frame, or the improvement of parameters describing the spatial structure without recourse to predictions from preceding frames, i.e. based solely on the spatial variation of gray or color values within this particular image frame.

2 The additional loop indicated in Fig. 4.6 by the medium and fat arrow with twin-roots emanating

- either from the pair synthetic and real image (fat)
- or from the pair view sketch and picture domain cues (medium).

This latter loop represents the updating process based on the differences between the prediction for frame time t+1 (synthetic image or view sketch) and the actual observations at frame time t+1 (recorded image or picture domain cues derived from it, respectively).

It should be noted that this approach deliberately represents an updating of the 3-D parameters, both for the structure of objects and illumination as well as for the relative motion between scene components and the camera, based on the 2-D differences in the picture domain. In this manner it is expected to take into account the measurement noise which appears to be modelled more easily in the 2-D picture domain where the original observations are taken. The fat arrow should indicate that the measurement errors are most easily taken into account if the raw image data are used directly. On the other hand, picture domain cues still have the advantage compared to 3-D estimates in that their derivation is closer to the original data and thus their error estimates involve only data from the current frame.

By now, it should be obvious that the latter loop (2) has been introduced in a manner which allows its quantitative realization by an Extended Kalman Filter approach. The extension of the framework introduced in Fig. 4.5 explicitly accommodates the machine vision approach developed by Dickmanns, Graefe and their group (see, e.g. Dickmanns and Graefe (1988)) for automatic driving of a van. The generic model of temporal developments required for such a process is indicated by the heavily shaded box across the intra-frame interpretation loops for each frame-time in conjunction with the broken arrow lines connecting the instantiated 3-D models (+) at frame-time t+1 to the predicted instantiations (-) at frame-time t+2.

Recently, Bajcsy (1988) pointed out that, for example, humans not only combine input from different sensor modalities like vision and touch, but in addition deliberately manipulate the parameters which influence the sensing process itself. Controlled variation of the sensing parameters allow one to disambiguate and simplify the interpretation of the input data obtained by the sensing process. She called this approach active perception. Specialized for machine vision, such an approach comprises a feedback loop from image interpretation to control of, e.g. location, attitude, focus, and stop of the camera. In order to accommodate such control loops explicitly in the framework of Fig. 4.5, modifications are necessary. Rather than discussing here such additional extensions concerning the sensing process, the remainder of this contribution is concerned with a discussion of refinements towards the representation of more complex activities.

5. THE REPRESENTATION OF MOTION CONTEXT AS A MEANS TO INFER INTENTIONS OF ACTORS

In an attempt to link observables like trajectories to motion verbs, Heinze et al. (1989) deliberately restrict themselves to motion verbs describing the motion of a car on private roads. In order to avoid an approach too much governed by ad-hoc considerations, criteria have been established for the selection of motion verbs (Cahn von Seelen, 1988; Heinze, 1989). Subsequently, a spelling dictionary for German (the "Duden") with about 150,000 entries has been scanned in order to extract all verbs given there, about 9,100. The selection criteria have then been used to collect those verbs (about 120) applicable to the discourse world of the recorded image sequences. Representations for these verbs have been designed with the goal to exploit them in order to map trajectories extracted from image sequences to the appropriate verbs. Preliminary results for an image sequence of several hundred frames indicate that the interpretation process selected automatically the appropriate 1 or 2 per cent of the verbs among those represented (Heinze, 1989; Heinze et al., 1989).

There is more, however, to the generic temporal descriptions than just trajectories: these descriptions should facilitate the representation of complex activities such as motions of deformable bodies, for example various gaits of human movement, from which even more complex descriptions can be derived, for example cooperative or antagonistic movements of entire groups of agents. Since movements are rarely free from plans and intentions, it is the aim to capture as much as is possible about the intentions of a mobile individual or a group of agents from their movements, be it facial movement, gesticulation, gait, or "body language" in even more general terms.

A tentative approach towards inferring gait from image sequences will be sketched here to indicate potential developments. Following

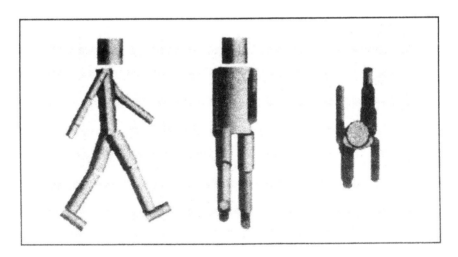

Fig. 4.7.   Coarse model of human, based on cylinders, proposed by Marr and Vaina (1980) in order to approximate a representation of human movements (from Hogg, 1983). Reprinted by permission of author and publisher.

Fig. 4.8. (a) Frame from an image sequence recording a walking human; (b) Projected edges of a coarse, cylinder-based model of a human superimposed on the image of Fig. 4.8a.

Hogg (1983), who applied ideas discussed by Marr and Vaina (1980), a coarse model of a human body with limbs is built from jointed cylinders—see Fig. 4.7.

Fig. 4.8a shows a frame from an image sequence recording a walking human, and Fig. 4.8b the projected edges of this coarse body model superimposed on the image of the walking human. The process of selecting this model, projecting it into the image plane of the recording camera, calculating the edge lines of the cylinders and selecting the projection and size parameters, such that these edges fit the image of the walking human, correspond to the right, down-directed part of the interpretation loop from Fig. 4.5. So far only components of the generic spatial description have been activated (activated in the sense that the discussion treats the spatial part. The temporal continuity and the transition to other gaits is treated in subsequent sections).

Fig. 4.9b shows an early series of photographs (taken in 1885) of a human slowing down from running to walking. In order to interpret an image sequence like this one, system internal representations of running, walking etc. have to be made available. Let us assume that representations for these human gaits, including an additional representation for a stationary pose, have been established as

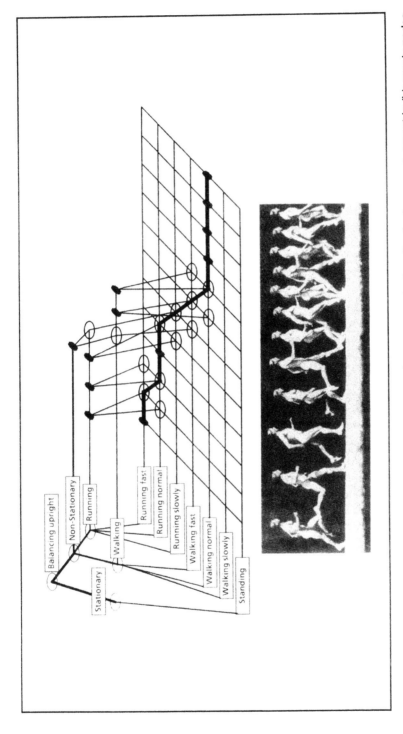

Fig. 4.9 (a) Conceptual hierarchy with respect to the motion verbs necessary to describe the human movement in (b); empty nodes are active, full nodes are dominantly active; (b) Chronophotography by E.-J. Mareys (1885) "Changes from running to walking".

subsequences of relative positions of torso and limbs. The following conceptual hierarchy will be used to illustrate the idea for recognizing the current gait:

- Balancing the human body on essentially straightened legs (root node deliberately chosen in order to exclude, for example, somebody kneeling with upright torso);

1. Stationary Pose
   - Standing
2. Non-Stationary Poses
   - Walking
     - Walking fast
     - Walking normally
     - Walking slowly
   - Running
     - Running fast
     - Running normally
     - Running slowly (Jogging)

Fig. 4.9a shows a hierarchical trellis of nodes. Each leaf node at the bottom of this hierarchical trellis represents a particular gait from the conceptual hierarchy, with the horizontal axis of the trellis corresponding to the temporal development as depicted in Fig. 4.9b. The nodes most likely corresponding to the human gait, depicted underneath in Fig. 4.9b, are activated, indicated here by ellipses. In transitional cases, more than one node may be activated at a frame time. Since this fact indicates an uncertainty in the association, the father node in the hierarchy is activated as well. If the transition occurs between concepts which are not siblings, even higher ancestor nodes are activated. This corresponds to an approach which chooses the most specific concept compatible with the observations. This concept node is indicated by a full ellipse and is denoted as dominantly active: (a node is dominantly active provided an instance of it is compatible with the input signals and none of its subnodes satisfies this requirement). In case of a switch between two nodes, neither of them has an instance fully compatible with the input signals, whereas a node higher up in the hierarchy—due to its less specific requirements—may still be compatible. Therefore, during transitions between different gaits, more general concept nodes are activated until one of the more specific nodes represents the temporal development again well enough to be selected as the dominantly active node, i.e. the currently best descriptor.

A generalization of this approach could be used to represent temporal developments not only of particular motion verbs, but of entire generically_describable_situations, abbreviated as gd_situations. Detailing ideas discussed in Nagel (1988), a gd_situation is conceived to capture not only the spatial structure and its temporal variability, but in addition the intentions of the agents involved. The schema given in Fig. 4.10 is suggested to represent the essential ingredients of a gd_situation. The introduction of explicit parameters indicating the prevalent spatial and temporal scale allows specification of the spatial as well as the temporal limits of a situation to be represented schematically. Moreover, it facilitates the distinction between variable and invariable components. The setting comprises all aspects which may appear as an influence to be taken into account. Among these, objects especially have to be represented with respect to their shape and material properties, in particular their surface properties. Since

```
                        Situation_Schema

  •  prevalent_spatial_scale; prevalent_temporal_scale ;
  •  Setting:
        Part considered to be invariable at prevalent_temporal_scale
        *   geometry ;
        *   material aspects of surrounding ;
        *   illumination ;
        *   non-material aspects of surrounding:
            - time_of_day ;
            - time_of_year ;
            - weather ;
            - sound ;
            - smell ;
        Part considered to be variable at prevalent_temporal_scale
        *   geometry ;
        *   material aspects of bodies (surfaces) ;
        *   illumination ;
  •  Reference_actors: set_of_actor(s) ;
  •  Associated_actors: set_of_actor(s) ;
  •  Boundary_Situation:spatial_input; spatial_output ;
                        temporal_input; temporal_output ;
  •  Temporal_links: set_of_preceding_situations:↑ situation_schema ;
                     set_of_following_situations:↑situation_schema ;
  •  Links_to_more_general_situations: ↑situation_schema ;
  •  Links_to_more_specific_situations: ↑situation_schema ;
```

Fig. 4.10.   Tentative components of a situation schema.

```
                       Actor_Schema
           •  identification ;
           •  ↑situation_schema ;
           •  Goal: set_of_plan(s) ;
                 plan: action_schema ;
           •  Activities: set_of_action
                 action: action_schema ;
```

Fig. 4.11.   Components of an actor-schema.

representations of these aspects are in principle available, they are not discussed here in more detail.

The essential aspect of modelling activities even more complex than movements is captured by the actor concept, see Fig. 4.11. This schema indicates that an actor will find himself always within at least one situation. Essential is the attempt to provide representations not only for the current activities of an actor, but also for his goals. Activities and goals are represented as a set of actions which in turn are both instantiations of the same action-schema—see Fig. 4.12. This should indicate that plans are considered to be potential actions, not (yet) currently being performed.

The representation for the bodily appearance of an actor is subsumed within the action schema where it will be part of the specifications of body components which are considered to be movable during an action. Particular activities are to be represented by a set of trajectories for all components involved, for example the set of trajectories of all limbs, head, and torso for the human model mentioned in connection with motion verbs. Although these schemata are still sketchy, they should

```
                      Action_Schema
          •  Configuration_of_actor_components:
             *  volumetric_description ;
             *  position ;
             *  attitude ;
             *  material_aspects ;
          •  Parameterized_component_trajectories ;
```

Fig. 4.12.   Components of an action-schema.

indicate the detail with which activities have to be represented. Preliminary results towards this goal are being investigated by Rohr (1989).

6. CONCLUSION

The interpretation of extended, image sequences recording complicated activities with multiple actors would consist in a search path through a hierarchical trellis of situation nodes in analogy to the example given for motion verbs in Fig. 4.9. This corresponds to a model based approach, where the notion of model comprises not only models for object bodies, but in addition models for abstract conceptual notions such as motion verbs or even more general concepts such as situations. Although it is known that a model-based approach towards the interpretation of a single image has so far encountered considerable problems due to the required computational expenses, there seems to be no way around it if the link to descriptions at a conceptual level has to be established.

Investigating these ideas leads to the necessity to circumscribe some of the notions associated with the natural language term "situation" more precisely (Baron, 1989). The setting may be decomposed into components which do not change and those which change during the recorded interval. In order to analyze the changes for nonstationary components, it has to be determined whether observable motions can be predicted using physical laws or whether additional degrees of freedom have to be taken into account. In the former case one may talk about a mobile object.

In the latter case, it will be important to delimit somehow the variety of possible temporal developments. For this purpose, the notion of intention will be attributed to visually perceptible components for the purpose of estimating the intentions as a means to predict observable motions. Such components have been denoted as "actors". If it becomes possible to infer the intention of an agent, this fact will allow making more reliable predictions of his future actions and thus of his discernible movements!

A stationary configuration of objects may be considered to provide a scenery. If an actor appears in the scenery, it is thereby converted into a "scene"; developments may occur, not all of which can be predicted merely on the basis of estimated physical state variables and physical laws governing their temporal change. Given a sufficiently comprehensive set of plans which are associated with characteristic movements of actors in the context of the observable scene, a task analogous to object recognition may consist in estimating the subset of plans compatible with observable movements. Each such plan may then be considered to correspond to a potential intention of the actor executing the observed movement. Intentions thus provide a potential means for predicting future observables as, for example, studied by Retz-Schmidt (1988), see also Andre et al. (1988).

The intentions of an actor depend on his actual model of the scene, including his actual model of other actors in the same scene, which implies his model of the intentions of other actors. Such a model of a scene derived by an actor is called his "situation". The situation of two actors in the same scene may be quite different, depending on the actual intentions of each actor and his model of the intentions of other actors in the scene. For example, one actor driving on a road may wish to overtake another one whereas this other actor does not wish to drive faster.

Representations for concepts mentioned here attempt to capture aspects which are usually discussed in connection with text and story understanding approaches. There are at least two major problems to be overcome in order to bridge the gap between the research areas of image and text understanding:

• Although the same words are used, the precise definition of the concepts underlying them is different, sometimes in a very subtle way. The notion of "situation" derived here from considerations of image sequence analysis, and the one discussed in-depth by Barwise and Perry (1983), starting from linguistics, logic and philosophy, may serve as an example. Moreover, it is acknowledged that:

the linguistic meaning of an expression in general greatly underdetermines its interpretation on a particular occasion of use (Barwise and Perry, 1983, pp. 36-37).

This fact shows that not only the concept for a word must be modeled, but in addition the entire context of its potential use to the extent it is admissible in the actual discourse world.

• The other problem concerns the necessity to model not only at the conceptual level, but also to provide explicit links between conceptual models for actions and intentions, and geometric models for the appearance and motion of the bodies involved.

Given these problems, it is clear that the gap is even greater if not only movements, but more complex behaviour has to be taken into account. Nevertheless, the link should now be apparent between the geometric models dominating animated computer graphics, as well as machine vision on the one hand, and the conceptual modeling in computational linguistics dominating approaches to understand texts.

ACKNOWLEDGMENTS

Critical remarks by W. Enkelmann, W. Krüger, and K. Rohr concerning a draft version of this contribution are gratefully acknowledged. I thank K. Rohr and C. Schnörr for providing the photographs of Fig. 4.8.

REFERENCES

J. Aloimonos and D. Shulman, "Integration of Visual Modules", Academic Press, Inc., Boston etc. 1989.

E. Andre, G. Herzog, and T. Rist, On the Simultaneous Interpretation of Real World Image Sequences and Their Natural Language Description: The System SOCCER. Proc. ECAI-88, 8th European Conference on Artificial Intelligence, München/FRG, 1-5 August 1988, pp. 449-454.

Anonymous, Proc. Workshop on Visual Motion, March 20-22, 1989, Irvine/California. IEEE Computer Society, ISBN 0-8186-1903-1.

L.L. Avant, A.A. Thieman, K.A. Brewer, and W.F. Woodman, "On the Earliest Perceptual Operations of Detecting and Recognizing Traffic Signs", in Vision in Vehicles, A.G. Gale, M.H. Freeman, C.M. Haslegrave, P. Smith, and S.P. Taylor (eds.), Elsevier Science Publishers B.V., Amsterdam, New York, Oxford, Tokyo 1986, pp. 77-86.

N. Ayache and F. Lustman, Fast and Reliable Passive Trinocular Stereovision. Proc. First International Conference on Computer Vision, London, June 8-11, 1987, pp. 422-427.

R. Bajcsy, Active Perception. Proc. IEEE 76 (1988) 996-1005.

H.P. Baron, Ein Beschreibungsansatz für Außenszenen auf dem Gelände des IITB. Diplomarbeit, Institut für Algorithmen and Kognitive Systeme, Fakultät für Informatik der Universität Karlsruhe (TH), Karlsruhe/FRG (Januar 1989).

J. Barwise and J. Perry, "Situations and Attitudes". The MIT Press, Cambridge/MA and London/England 1983.

R.C. Bolles, "Verification vision for programmable assembly", Proc. Int. Joint Conf. Artificial Intelligence, Cambridge/MA, 22-25 August 1977, pp. 569-575.

O.J. Braddick and A.C. Sleigh (eds.), "Physical and Biological Processing of Images", Springer-Verlag Berlin Heidelberg New York 1983.

M. Brady, Seeds of Perception. Proc. Third Alvey Vision Conference, 15-17 September 1987, Cambridge/UK, pp. 259-265.

U. Cahn von Seelen, Ein Formalismus zur Beschreibung von Bewegungsverben mit Hilfe von Trajektorien. Diplomarbeit, Fakultät für Informatik der Universität Karlsruhe (TH), August 1988, Karlsruhe.

J Canny, A Computational Approach to Edge Detection. IEEE Trans. Pattern Analysis and Machine Intelligence PAMI-8 (1986) 679-698.

R. Deriche, Using Canny's Criteria to Derive a Recursively Implemented Optimal Edge Detector. Internal Journal of Computer Vision 1 (1987) 167-187.

E.D. Dickmanns and V. Graefe, Applications of Dynamic Monocular Machine Vision. Machine Vision and Applications 1 (1988) 241-261.

O.D. Faugeras, A Few Steps towards Artificial 3D Vision. In "Robotics Science", M. Brady (ed.), MIT Press, Cambridge, Mass., and London England, 1989. pp.39-137.

J.A. Feldman, "Four Frames Suffice: A Provisional Model of Vision and Space", The Behavioral and Brain Sciences 8 (1985) 265-289.

M.M. Fleck, Representing Space for Practical Reasoning, Proc. Tenth International Joint Conference on Artificial Intelligence IJCAI-87, 23-28 August 1987, Milan/Italy, pp. 728-730; see, too, Image and Vision Computing 6 (1988) 75-86.

A. Hanson and E. Riseman, "The VISIONS Image-Understanding System", in Advances in Computer Vision vol. 1, Brown C. (ed.), Lawrence Erlbaum Associates, Inc., Hillsdale/NJ 1988, pp. 1-114.

G. Healey and T.O. Binford, Local Shape from Specularity. Proc. First International Conference on Computer Vision, London, June 8-11, 1987, pp. 151-160.

N. Heinze, Realisierung eines Ansatzes zur Berechnung von Bewegungsverben aus Trajektorien. Diplomarbeit, Fakultät für Informatik der Universität Karlsruhe (TH), April 1989, Karlsruhe.

N. Heinze, W. Krüger, H.H. Nagel, Zuordnung von Bewegungsverben zu Trajektorien in Bildfolgen von Straßenverkehrsszenen. Proc. DAGM-Symposium 1989, Informatik-Fachberichte 219, Springer-Verlag Berlin Heidelberg New York 1989, pp. 310-317.

D. Hogg, "Model-based Vision: A Program to See a Walking Person", Image and Vision Computing 1 (1983) 5-20.

B.K.P. Horn, "Robot Vision". The MIT Press, Cambridge/MA 1986.

B.K.P. Horn and M.J. Brooks, The Variational Approach to Shape from Shading. Computer Vision, Graphics, and Image Processing 33 (1986) 174-208.

B.K.P. Horn and M.J. Brooks (eds.), "Shape from Shading". The MIT Press, Cambridge/MA and London/England 1989.

T. Kanade, "Region segmentation: signal versus semantics", Proc. Int. Joint Conf. Pattern Recognition, Kyoto/Japan, 7-10 November 1978, pp. 95-105; see also Computer Graphics and Image Processing 13 (1980) 279-297.

K.I. Kanatani and T.C. Chou, Shape from Texture: General Principle. Artificial Intelligence 38 (1989) 1-48.

G.J. Klinker, S.A. Shafer, and T. Kanade, The Measurement of Highlights in Color Images, International Journal of Computer Vision 2 (1988) 7-32.

E. Krotkov, Focusing. International Journal of Computer Vision 1 (1987) 223-237.

M.D. Levine, "Vision in Man and Machine", McGraw-Hill Book Company, New York 1985.

H. Li and J.R. Kender (eds.), Special Issue on Computer Vision, Proc. IEEE 76 (1988) 859-1050.

D. Marr, "Vision", W.H. Freeman and Company, San Francisco/CA 1982.

D. Marr and L. Vaina, "Representation and Recognition of the Movements of Shapes", AI-Memo 597 (October 1980), Artificial Intelligence Laboratory, MIT, Cambridge/MA.

H.H. Nagel, "Über die Repräsentation von Wissen zur Auswertung von Bildern", in Angewandte Szenenanalyse, J.P. Foith (ed.), Informatik-Fachberichte 20, Springer-Verlag Berlin Heidelberg New York 1979, pp. 3-21.

H.H. Nagel, "Image Sequences—Ten (Octal) Years—From Phenomenology towards a Theoretical Foundation", Proc. 8th International Conference on Pattern Recognition, Paris/France, 27-31 October 1986, pp. 1174-1185.

H.H. Nagel, "Principles of (Low-Level) Computer Vision", in Fundamentals in Computer Understanding: Speech and Vision, J.P. Haton (ed.), Cambridge University Press, Cambridge/UK 1987, pp. 113-139.

H.H. Nagel, "On the Estimation of Optical Flow: Relations between Different Approaches and Some New Results", Artificial Intelligence 33 (1987) 299-324.

H.H. Nagel, "From Image Sequences towards Conceptual Descriptions", Image and Vision Computing 6 (1988) 59-74.

V.S. Nalwa and T.O. Binford, On Detecting Edges. IEEE Trans. Pattern Analysis and Machine Intelligence PAMI-8 (1986) 699-714.

V.S. Nalwa, Edge-Detector Resolution Improvement by Image Interpolation. IEEE Trans. Pattern Analysis and Machine Intelligence PAMI-9 (1987) 446-451.

G. Retz-Schmidt, A REPLAI of SOCCER: Recognizing Intentions in the Domain of Soccer Games. Proc. ECAI-88, 8th European Conference on Artificial Intelligence, München/FRG, 1-5 August 1988, pp. 455-457.

K. Rohr, Auf dem Wege zu modellgestütztem Erkennen von bewegten nicht-starren Körpern in Realweltbildfolgen. Proc. DAGM-Symposium, Mustererkenung 1989, Informatik-Fachberichte 219. H. Burkhardt, K.H. Höhne, B. Neumann (eds.), Springer-Verlag Berlin Heidelberg New York 1989, pp. 324-328.

K. Sugihara, Machine Interpretation of Line Drawings. The MIT Press, Cambridge/MA and London/England 1986.

CHAPTER 5

ProtoKEW: A knowledge-based system for knowledge acquisition

Han Reichgelt and Nigel Shadbolt
Artificial Intelligence Group, Department of Psychology,
University of Nottingham, Nottingham NG7 2RD

ABSTRACT

In this paper we describe a system aimed at providing software support for the process of knowledge acquisition. Such support comprises a workbench incorporating a number of knowledge acquisition tools. These include knowledge elicitation techniques such as sorting and rating methods, together with machine learning techniques. The paper discusses the various problems raised by this work. These include: defining an adequate view of the general acquisition process, developing an appropriate implementation architecture, directing knowledge acquisition via knowledge level models and producing a sufficiently powerful representation language to integrate the results of acquisition. Finally, we describe the limitations of our current system and propose future developments in our work.

1. INTRODUCTION

The acquisition of knowledge remains the critical phase of expert system development. Recently a number of software support tools have appeared that provide for the analysis, refinement and integration of knowledge from diverse sources: texts, manuals, verbal transcripts, cases, and the experts themselves.

Consideration of these systems reveals three broad classes of support tool (Shadbolt and Wielinga, 1990). One class provides support for acquisition in specific domains or else allows acquisition in domains that share particular problem solving strategies. A second consist of

171

computer implementations of particular knowledge acquisition techniques and are as such domain independent. The third class we might refer to as *loosely integrated* systems.

Illustrative of the first class of support tools is the SALT system (Marcus 1988) which has built-in knowledge of the problem solving strategies used when configuring electro-mechanical devices. The problem solving strategies are variants on a propose-and-revise method and as such SALT can be used in applications which share this problem solving method. The expert interacts through an automated, highly structured, interview which allows the selection by menu of elements in the domain. The interview results are converted into rules by SALT and the expert then has the opportunity of refining these. A different approach is taken by Friedland (1979). In his system, expert geneticists code up the knowledge base themselves in a language customised to genetics.

In the second class of support tools we see a number of software products that implement specific acquisition techniques. One of the first to be supported was the repertory grid technique which we will describe later in this paper. It is time consuming and laborious to apply manually, but the underlying algorithms are straightforward. It was inevitable that it should be automated. For example, Boose (1985) developed the Expertise Transfer System (ETS). ETS and its variants have been used in a great many applications and domains (see Boose, 1986 for a list). In this paper we describe two other implementations of standard manual elicitation techniques- laddering and card sorts.

Each of these individual techniques is limited in scope to the elicitation of certain types of knowledge. Moreover, each displays its own advantages and drawbacks. This also applies to the other category of acquisition tools, namely machine learning algorithms. We have incorporated one of the most commonly encountered, Michalski's AQ11, which induces rules in the form of structural descriptions that discriminate positive and negative examples of a concept. AQ11 belongs to the family of similarity based learning techniques (SBL). Experience with ML techniques has shown that they are very sensitive to the quality and quantity of examples used as the learning set. Thus, they can be affected by the errors in data and the particular description of a case. The lesson to draw is that used alone any individual knowledge acquisition method is unlikely to provide sufficient coverage of any application domain.

The final class of support tools covers a range of systems that attempt to provide a set of acquisition functions. Examples of such software include KADS Power Tools (Anjewierden, 1987) and KEATS (Motta, Rajan, Domingue and Eisenstadt, 1990). These combine a number of documentation and browsing aids to help the knowledge engineer find his way around the acquisition material. They also include a few particular elicitation methods. What they lack is any significant integration.

Each type of support system is restricted in scope. The aim of the ESPRIT II Project (P2576) (Note 1). ACKnowledge is to achieve integration between a wide range of acquisition techniques, and to combine the best features of current support tools. That is, the goal is the construction of a knowledge engineering workbench (KEW). KEW would implement a range of elicitation techniques, incorporate machine learning methods, be applicable across domains, and embody a principled or knowledgeable approach to the entire acquisition process. To build such a workbench would require that we effectively build a knowledge-based system for knowledge engineering.

This view has important consequences for the next generation of knowledge acquisition tools. As argued in Shadbolt and Wielinga (1990), the most important outcome is that such a workbench will have to be an active and dynamic system. It will be a system animated by a variety of knowledge sources, including knowledge about: conducting knowledge acquisition, using various knowledge acquisition tools, integrating the results of acquisition into a consistent and changing application knowledge base. We also expect to exploit synergies between techniques within an integrated framework.

As part of the ACKnowledge work a number of conceptual prototypes of such knowledge engineering workbenches have been developed. In this paper we describe ProtoKEW, a system developed at the University of Nottingham.

The structure of this paper is as follows. In the next section we present the view which is adopted within the ACKnowledge project of the knowledge acquisition process. In Section 3 we describe an architecture to implement the concept of an integrated and active acquisition workbench. Section 4 shows how control of the knowledge acquisition process is effected within this architecture. Section 5 describes the underlying knowledge representation language that allows for the integration of results in the workbench. Section 6 describes the subset of tools chosen for inclusion in ProtoKEW, together with our reasons for choosing this particular set. Section 7 examines in detail how ProtoKEW translates the outputs of the various tools into the common underlying knowledge representation language. Section 8 describes the problems of integrating these results into a consistent knowledge-based system. Section 9 discusses the issues that emerge when evaluating the products of acquisition. Finally, section 10 discusses the substantial issues that remain to be investigated, and summarises our experience in building ProtoKEW.

2. THE KNOWLEDGE ACQUISITION PROCESS

In this section we detail the assumptions we make about the various ingredients and processes that have to be included in any automated knowledge acquisition workbench.

There are two approaches in the literature to the problem of knowledge acquisition. The first is the rapid prototyping approach. In this approach, one constructs a running prototype system as soon as possible. Further acquisition then involves debugging the prototype with the help of the domain expert, possibly leading to a more radical re-implementation.

The second approach, and the one taken within our project, is a modelling approach. In this view, acquisition is conceived as both a *constructive* and *interpretive* process. Under this view, acquisition cannot be regarded as simply the *transfer* of knowledge from one entity to another. In any acquisition process one attempts to *model* the application domain and the associated problem solving expertise. A model of an object, device, process or procedure is an abstraction of selected characteristics. The criteria for which characteristics at what level of detail to use in the model are subjectively determined. They are determined by the intentions and purposes which users and designers of the system have in mind.

Such models are clearly at the *knowledge level* in the sense that Newell (1982) uses the term. However, they say nothing about how this knowledge might be implemented in a run-time program. Within our work we follow the KADS (Breuker and Wielinga, 1987, 1989) approach of distinguishing four knowledge layers at the knowledge level. The first of these is *domain* knowledge and describes the domain concepts, elements and their relations. A second type is *task* knowledge, which has to do with how goals and sub-goals, tasks and sub-tasks should be performed. A third sort of knowledge is referred to as *strategic*. This is used to monitor and control problem solving. Finally, we distinguish *inference layer* knowledge. This has to do with how the components of problem solving and expertise are to be organised and used in the overall system. The inference layer is sometimes used to discriminate different classes of expert system; for example, diagnostic from planning systems.

A considerable amount of current research attempts to understand expertise in terms of models of various knowledge layers: problem solving, task models and domain representations (Steels, 1990). The hope is that by understanding the content and form of these models we can use them to inform acquisition. For example, Breuker and Wielinga (1989) advocate the use of models of problem solving. These so-called

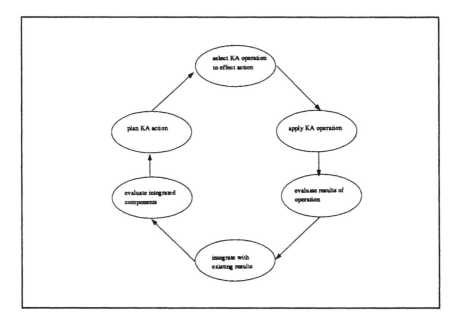

Fig. 5.1.   The knowledge acquisition activity cycle.

interpretation models are knowledge constructs at the inference layer—they indicate how components of expertise are organised and used in problem solving. An alternative approach is to build task layer representations (Bylander and Chandrasekaran, 1988; Clancey 1985). Here the aim is to use general, invariant features of a task to guide KA. Finally, a number of researchers exploit the fact that many KBS applications are built in similar, if not nearly identical domains (Eshelman 1989; Marcus, 1988). In this case one attempts to build reusable domain descriptions. We believe that ultimately models and structures at all of these levels will be important in any informed and directed knowledge acquisition approach. In ProtoKEW we illustrate the use of models which originated in both the work of Breuker and Wielinga, and Clancey.

Knowledge acquisition as a process can be seen to have a cyclic structure, as illustrated in Fig. 5.1. Data are collected, interpreted, integrated with existing knowledge, and plans for a new cycle are made. This process can occur at different phases of the life cycle of KBS development and maintenance. Moreover, the emphasis on any aspect of the process may vary.

In order to support this activity, our KBS for knowledge acquisition would need a variety of knowledge sources. A first type of knowledge has to do with knowledge about the knowledge acquisition process itself. This would be advice and guidance about what to do when. Moreover, it would need to be sensitive to where one is in the activity cycle in Fig. 5.1.

A second source of knowledge is the use of models in the broadest sense of the term. This could be knowledge resident in the system about classes of problem solving, epistemological categories, domain structures and so on.

Knowledge about the knowledge acquisition tools themselves constitute a third type of knowledge to be used in any KBS for knowledge acquisition. Thus tools make assumptions about data, how it is represented and analysed. The system and knowledge engineer should have explicit command and use of this knowledge.

A fourth source of knowledge has to do with our theory and account of how to integrate knowledge acquired from different tools into a consistent evolving application knowledge base. The system will need to have methods for transforming between the knowledge representation languages of the tools and of the underlying system, as well as integrating this transformed knowledge.

Finally the emerging application KB needs to be evaluated. Our system needs knowledge about the analysis of accrued and possibly integrated knowledge, including simple analysis for consistency, coverage and granularity.

Each of these types of knowledge is important in the implementation of the knowledge acquisition KBS we envisage. In the sections that follow we attempt to show how these various components feature in ProtoKEW.

3. ARCHITECTURE FOR PROTOKEW

ProtoKEW has been implemented in CommonLisp and PCE (Anjewierden and Wielemaker, 1989), a high-level object-oriented graphics package. In this section we describe the global architecture of ProtoKEW. Several aspects of the architecture are discussed in greater detail in later sections. Fig. 5.2 shows the architecture of ProtoKEW.

ProtoKEW makes a strict separation between what one might call "Knowledge Engineering Knowledge" and "Application Knowledge". The former refers to knowledge about knowledge engineering in general, and the particular steps that the knowledge engineer takes for building an application. "Application Knowledge" refers to knowledge relevant to the application, that has been acquired either from an expert or by some other means. The Knowledge Engineering Knowledge is stored in the Knowledge Engineering Knowledge Base (KEKB), while Application

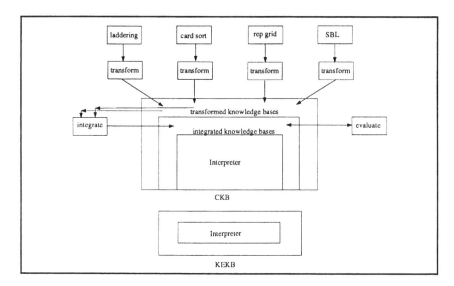

Fig. 5.2.   The architecture of ProtoKEW.

Knowledge is stored in the Common Knowledge Base (CKB). Both knowledge bases have their own associated interpreters.

The KEKB contains the knowledge necessary to drive the knowledge acquisition process. ProtoKEW is an interactive system that can guide the user through the knowledge acquisition process, and provide advice when necessary. However, ProtoKEW will not take any actions without being explicitly instructed to do so by the user. In section 4, we discuss this role of ProtoKEW and how the KEKB supports this process in more detail.

ProtoKEW contains a number of knowledge acquisition tools. Each tool typically uses its own knowledge representation language. The reasons for this are both pragmatic and theoretical. Pragmatics dictates that rather than building all tools from scratch we incorporate a number of existing tools. In section 6 we describe the kinds of theoretical considerations that led us to allocate tools their own KRLs.

Since the different tools in ProtoKEW can use different knowledge representation languages, ProtoKEW must allow for the transformation of knowledge from one formalism into the formalism used by the CKB. ProtoKEW contains a number of transformation tools. As we shall see in section 7, for some knowledge representation languages, transformation turns out to be impossible without further consultation with the domain expert. We regard the interaction with the domain during transformation as an important extra phase of acquisition. We will refer to the outcome of the transformation process as a "Transformed Knowledge Base" (TKB).

Transformed knowledge bases are stored in the CKB. However, simply transforming tool-specific knowledge bases into the same knowledge representation language is not sufficient. One of the assumptions underlying ProtoKEW is the belief that integrating the results of different knowledge acquisition tools produces a more powerful knowledge base than any produced by a single knowledge acquisition method. ProtoKEW therefore contains a knowledge integration tool; knowledge integration is discussed in more detail in section 8.

Finally, evaluation and validation of the knowledge both in tool-specific knowledge bases and in the CKB must be supported within our workbench. Such evaluation will consist of both logical and operational, semantic and pragmatic tests. These are discussed in section 9.

4.   CONTROL IN PROTOKEW

As discussed in section 2 we are aiming to build an active and directive knowledge acquisition workbench. The evolution of such a system will be a gradual process. Initially we will use the knowledge engineer as the main controller of KEKB activity. Ultimately we hope to encapsulate more and more knowledge about knowledge acquisition in the KEKB. At this early stage in ProtoKEW's development we have a mixed initiative system with control very much in the hands of the knowledge engineer. The knowledge in the KEKB is currently restricted to a number of compiled knowledge structures and very little active reasoning about the goals and activities to pursue takes place.

Fig. 5.3 shows the control interface to ProtoKEW.

Control is exercised in the upper half of the display whilst the individual acquisition tools carry out their interactions with the user in the lower half. The control panel is a set of pull down menus and display windows. The first of these in the top left hand corner is used to display the *directive model* under consideration in the acquisition process. In this case it is displaying a library object resident in the KEKB which is the interpretation model for heuristic classification (Clancey, 1985). We assume that selection of this object has been made by the knowledge engineer. It serves as a candidate for the type of problem solving observed in the application domain (Note 2). The point to note about selection of this structure is that it sets the context for subsequent acquisition.

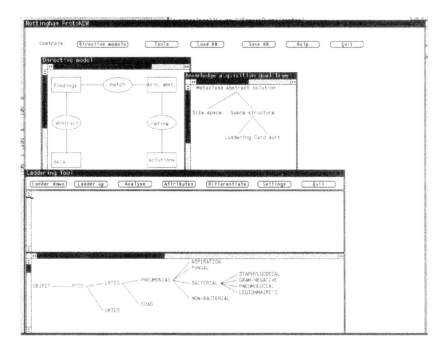

Fig. 5.3.   Control Interface to ProtoKEW.

The directive model contains knowledge which can be used to inform acquisition. It shows us both the inputs and outputs of various processes that make up a generic type of problem solving. There are four kinds of input/output classes, also called meta-classes, for this model structure: data, findings, abstract solutions and solutions. The processes that operate on these meta-classes are: abstract, match and refine. Each of these processes is in turn associated with a set of methods for effecting the change from input to output. Thus, refine is defined as a process of refining solution abstractions into more specific solutions; a method which can effect this is specialisation.

It is precisely this sort of knowledge that can generate acquisition goals. By looking more deeply into the heuristic classification structure we can establish goals and actions to take. In Fig. 5.3 we see that the abstract solution meta-class has been selected and this has activated knowledge in the KEKB, which suggests how knowledge might be acquired about this part of the expertise space. This is shown as a knowledge acquisition goal tree in the window to the right of the directive model.

The goal tree indicates that in this context we can use either laddering or a card sort to explore the structure of the solution space. It is this knowledge that leads the knowledge engineer to select an option from the tools menu. We can see in Fig. 5.3 that the laddering elicitation tool has been selected and is actually present in the tool window. Should the knowledge engineer need extra information he can select objects on the knowledge acquisition goal tree. Thus selecting the laddering node would cause ProtoKEW to produce a set of goals and actions to take in such a session.

5.   THE COMMON KNOWLEDGE REPRESENTATION LANGUAGE

One of the central components in the architecture of any Knowledge Engineering Workbench is the module that stores the knowledge acquired. In section 3, we referred to this component as the Common Knowledge Base (CKB). We begged the question as to which knowledge representation language to use for this component.

There is a trend among those building KEWs to use hybrid knowledge representation languages for the CKB (e.g. Gaines and Linster, 1990). Elsewhere, it has been argued that whilst such hybrid systems are extremely useful as programming environments, their use as knowledge representation tools is limited (Jackson, Reichgelt and van Harmelen, 1989, p. 3). In particular, since these systems give no guidelines as to when to use which of the different component representation languages, the knowledge engineer is often faced with a bewildering array of possibilities. As a result, they encourage an *ad hoc* style of programming and representation. This problem is even worse in the present context, where integration of knowledge from different sources is of considerable importance. If knowledge is to be integrated, then it becomes important to identify, for a given piece of knowledge, contradictory information. If such information can be stored in a large number of different ways, then finding contradictory knowledge becomes at best tedious, and at worst intractable.

Because of the drawbacks of using hybrid systems, ProtoKEW uses logic as the knowledge representation language for the CKB. The arguments in favour of using logic as a knowledge representation language are well-known (see, e.g., Hayes, 1977; Moore, 1984), and include the fact that logics have a model-theoretic semantics which allows one to determine the correctness (soundness) of the interpreter, and the fact that logic provides unsurpassed expressive power. Moreover, in the present context, the use of logic as the Common Knowledge Representation Language has the added advantage that it makes the problem of transformation and integration considerably easier; a point to which we return at some length in sections 7 and 8.

The particular theorem prover that we are using is an implementation of a semantic tableaux proof theory for classical first-order predicate calculus. The background theory for such a system is explained in detail in Elfrink and Reichgelt (1989). However, we have improved over this specification in a number of ways. First, Elfrink and Reichgelt describe only a backward chainer; the variant used here can also be used in a forward chaining fashion.

Second, the theorem prover described by Elfrink and Reichgelt contains hard-wired control heuristics. The theorem prover used in ProtoKEW's CKB allows the user to change the heuristics used for conflict resolution, conjunct ordering, and backtracking. This provides control over the way in which the proof space is searched.

In ProtoKEW the CKB contains a partitioning mechanism which allows one to divide the logical axioms into different partitions. In logical terms, each partition corresponds to a different theory. While the system attempts to ensure consistency within a single partition, it does not ensure consistency between partitions. The partitioning mechanism therefore allows users to explore different theories simultaneously.

ProtoKEW provides a number of functions for adding a proposition to a partition in the CKB. These functions differ in the amount of testing that they do before a proposition is added to the CKB. There are basically two types of test, a redundancy test and a consistency test. The redundancy test determines whether a proposition is already entailed by the existing propositions in the partition. If a proposition fails this test, then it is not added to the knowledge base. The consistency test involves determining whether the proposition is consistent with the other propositions stored in the partition. Can the system prove its negation from the information already in the knowledge base? If a proposition fails this test, then the reason maintenance system is called on to ensure consistency. We discuss this in more detail in the section on integration, section 8. If *both tests* are performed successfully, then the user is guaranteed that the set of propositions is minimal and consistent, at least within the limitations of the theorem prover (Note 3). Performing both the redundancy and the consistency test between partitions is a relatively slow process. ProtoKEW therefore allows the user to forego either of the tests. The risk is, of course, an integrated knowledge base lacking one or both of the properties of minimality and consistency.

The theorem prover has been interfaced to a McAllester style reason maintenance system (RMS) (McAllester, 1980). Before a theorem prover uses a particular proposition in its search for a proof, it will first consult the RMS to ensure that the proposition has support-status "in" (believed to be true). Knowledge acquisition is often an iterative process in which previously accepted pieces of knowledge have to be rejected as more information becomes available, or as the expert changes his or her mind. We believe that this can only be achieved using an integrated theorem prover and RMS. In addition, an RMS turns out to be extremely useful when integrating knowledge from different sources. We return to this aspect of the CKB in section 8.

6. KNOWLEDGE ACQUISITION TOOLS IN PROTOKEW

Four knowledge acquisition tools were chosen for inclusion in the ProtoKEW: ALTO, an implementation of the laddered grid technique, a card sort program, an implementation of the repertory grid method, and a Similarity Based machine learning algorithm. Each of the tools will be described in more detail later in this section.

The techniques were selected for the following reasons. Each was available to us. Each used a well-defined method of acquisition. Together they allowed for a range of different types of knowledge to be acquired. Finally, it was felt that it would be possible to understand the various ways in which the input and output were being interpreted by the user of ProtoKEW.

The ALTO system (Major and Reichgelt, 1990) implements the laddered grid method (see e.g. Shadbolt and Burton (1990)). In laddering the expert and the knowledge engineer construct a graphical representation of domain knowledge in terms of the relations between knowledge elements. The result is a two-dimensional graph whose nodes are connected by labelled arcs. In using the technique the elicitor enters the conceptual map, and then attempts to move around it with the expert.

The laddering method can be used to elicit various types of knowledge. The use which we will illustrate is domain oriented laddering. Here the emphasis is on what Clancey (1983) calls structural knowledge—knowledge about the types of entity present at the domain level.

Typically laddering involves the following steps:

1. Ask the expert to generate a starting point (seed item) and start him or her off with this seed
2. Move around the domain using the following prompts:
   a. To move down the expert's domain knowledge
      Can you give an example of <item>?

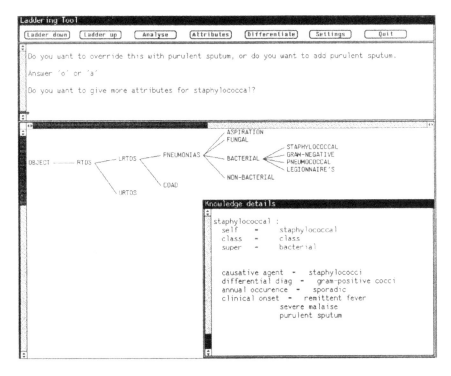

Fig. 5.4.   Laddering in the domain of respiratory diseases.

    b. To move across the expert's domain knowledge
        What alternative examples of <superclass> of <item>
        are there?
    c. To move up the expert's domain knowledge
        What are <same level items> examples of?
  3. Elicit attributes for the various concepts
        What have <same level items> got in common?
        What is the key difference between <item1> and <item2>?

In use, one may move around the domain using these prompts in any order that the expert or knowledge engineer find convenient. A number of studies have revealed that laddering is a powerful elicitation techniques (Burton, Shadbolt, Hedgecock and Rugg, 1987; Burton, Shadbolt, Rugg and Hedgecock, 1988; Shadbolt and Burton, 1989), which yields large amounts of knowledge in a cost-effective way.

Fig. 5.4 shows an ALTO screen from the domain of respiratory diseases. ALTO uses a simple object-oriented language as its tool-specific knowledge representation language, CommonSloop

(Reichgelt, Major and Jackson., 1990). Each node in the laddered hierarchy is represented as an object in CommonSloop. An object can have user-defined slots which are properties of the item in the node hierarchy. CommonSloop supports multiple inheritance. Inheritance also follows a default principle; attributes associated with higher-level objects are inherited by their children but can be overridden in the children.

The use of the object-oriented style of representation for laddering is both natural and intuitive. In structured object representations we aim to bring together under a simple indexing method the associated properties of objects. The use of inheritance allows us to exploit taxonomic relations in an efficient way. Specialising objects by overwriting or adding attributes and their values provides for a means of discriminating between objects.

As we shall see in section 7 the choice of this style of representation for ALTO has important consequences for the way in which the results of ALTO are transformed into the CKRL of ProtoKEW.

The second tool to be incorporated was the card sort. This technique is useful when one wishes to uncover the different ways an expert sees

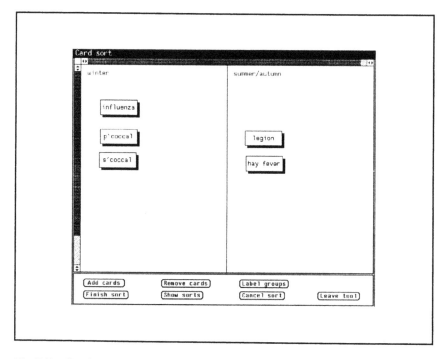

Fig. 5.5.   Card sort in the domain of respiratory diseases.

relationships between a set of concepts. The version we have automated allows the user to generate a set of cards each of which he labels with a concept word. The expert is asked to sort the cards into any number of categories he finds appropriate. Each category is then labelled. The expert is also asked to provide a name for the criterion on which the sort is based. This process is repeated as long as the expert is able to suggest further categories.

Fig. 5.5 shows a screen of a sort, once again, from the domain of respiratory diseases.

When using this technique on different occasions we can direct the expert to generate cards of very different epistemological categories. In the sort above we see that the cards name specific diseases, and that the sort is in terms of when they typically occur. However, the expert might be sorting cards which name symptoms or observed data; the categories might be the types of patients the symptoms occur in. The actual sort categories themselves may be used as the elements for other sorts, giving us so called hierarchical sorts. It is important to understand that the way in which a tool is used, what kind of knowledge is being sought, will vary depending on the context of acquisition.

The third tool chosen for inclusion is the repertory grid technique which has its roots in the psychology of personality (Kelly, 1955). Like the card sort it is designed to reveal a conceptual map of an area or aspect of expertise. However, in a number of respects this technique engages in a greater degree of analysis than is possible with the standard card sort.

The expert is once again given, or asked to generate, a set of elements. He is then asked to chose three, such that two are similar, and different from the third. Thus the expert might be presented with gram negative pneumonia, legionnaire's disease, and staphylococcal pneumonia. Legionnaire's is different from the other two since it involves gastro-intestinal upset. The dimension by which these elements have been separated is called the construct. The ends of the construct are called poles and the construct is taken to be measurable, that is we assume that elements can be rated along the construct.

This process is continued with different triads of elements until the expert can think of no further discriminating constructs. The result is a matrix relating elements and constructs. The matrix is analysed using a statistical techniques called cluster analysis. This reveals clusters of elements and constructs which represent meaningful concepts. These concepts may not have been articulated in the original elicitation session. We are also able to subject the results of cluster analysis to the technique of entailment analysis (Gaines and Shaw, 1986). This latter analysis allows us to generate implications between constructs. For example, we might find that the pole *severe obstruction* of the construct *obstructive ventilatory defects* implies *low FEV1/FVC ratio* on the construct *FEV1/FVC ratio*. Fig. 5.6 shows our repertory grid tool.

As with card sorting it is possible to rate elements representing very different sorts of epistemological category.

The final traditional acquisition tool included in ProtoKEW is an implementation of AQ11 (Michalski and Larson, 1978). This machine learning algorithm seeks to find the most general rule that describes training instances. The method seeks the most general form of a rule that discriminates one class from all other classes. We can think of the rules describing classes that we are trying to induce as comprising a space we are trying to search, the *version space* (Mitchell, 1979). At one extreme is the null description in which all conditions on the rules have been dropped and which describes anything. The most specific parts of the rule space are the individual training instances; these simply describe individual instances of classes. We can visualise this space as shown in Fig. 5.7. This space is divided into sets: G the set of most general rules (hypotheses, concept descriptions or discriminators), S the set of most specific rules. One is attempting to minimize the set H so that one has the most general rules sufficient to discriminate a class given the specific training set.

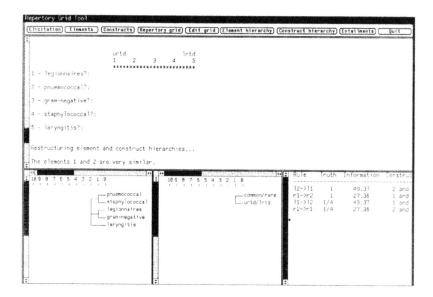

Fig. 5.6.   Repertory grid in the domain of respiratory diseases.

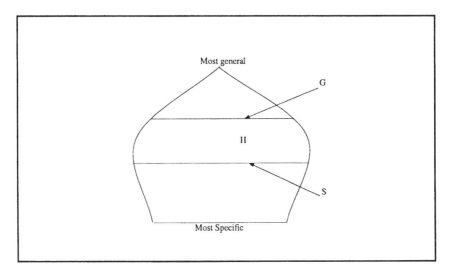

Fig. 5.7.   The version space using the G and S boundary sets.

Initially, the version space is the complete rule space of possible concept descriptions. As training examples are encountered, candidate concepts are eliminated from the version space. If the instance is a positive example, then we first remove from G any rules that do not cover the new example; we next generalise the elements in S as little as possible so that they will cover this new positive example. If the instance is negative, then one first removes from S all concepts that cover this counter example; one then specialises the elements in G as little as possible so that they will not cover this new negative example. It will be apparent that candidate elimination is a conservative least-commitment strategy—it does not eliminate unless forced to do so. Given an appropriate set of training instances, the technique will converge to a single concept in H.

A problem with this approach is that of learning disjunctive concepts. The least generalisation of a positive example will always be a disjunction of S with the new training example. In candidate elimination the introduction of disjunctions is a way of avoiding generalisation.

The solution adopted in AQ is to perform repeated candidate-elimination runs to find several conjunctive descriptions that together cover all of the training instances. We repeatedly find a conjunctive concept description that is consistent with some of the positive training instances and with none of the negative ones. The positive instances that have been accounted for are removed from further consideration. The process is repeated until all positive instances have been covered.

One can make use of an additional heuristic that selects a seed, a positive training instance that has not been covered by any description in G. This has the effect of choosing training instances that are far apart in the instance space.

7. TRANSFORMATION

Each of the knowledge acquisition tools described in the previous section use their own representation languages. In order to achieve the hoped-for synergy effects, the knowledge produced by the different tools needs to be transformed into the Common Knowledge Representation Language.

ProtoKEW contains different transformation algorithms for each of the four tools already discussed. We will now describe the various transformation algorithms in some detail.

Knowledge acquisition tools embody implicit assumptions about the type of knowledge to be acquired. The transformation algorithms rely on these assumptions. As indicated in section 6 our various acquisition tools could in principle be used to elicit different types of knowledge. In particular, we allow each of our tools to operate at each of the four knowledge levels discussed in section 2. Thus, in principle, we could ladder the structural concepts of the domain level, or in another instance ladder the structure of the task level. However, in our present implementation we make the assumption that acquisition always takes place at the domain level, and in the rest of this section we will frame our discussions accordingly.

The knowledge acquired in a laddering tool will concern objects in the domain and sets of such objects and their attributes. A card sort would presuppose that we have a set of objects in the domain, or a set of classes of such objects, and then acquires similarities and differences between the objects or classes. A repertory grid would presuppose a set of elements and then acquire properties that these elements may have. If the repertory grid allows one to do an entailment analysis, then it will produce implications to the effect that a value range on one property entails a value range on another property. An AQ11 algorithm will start with a description of a set of instances of classes, where each instance is described in terms of their attributes. It attempts to produce the most general descriptions that describe the relevant classes.

Each tool uses a representation for its output knowledge appropriate to the form of knowledge produced. For example, we have seen that ALTO stores its output in CommonSloop. AQ11 generates discriminations hierarchies.

In transforming tool-specific knowledge bases into the CKRL, the assumptions underlying both the use of a tool and the tool-specific knowledge representation language are made explicit. Sometimes, this can be done automatically; sometimes, this will involve interaction with the user.

Consider a knowledge base produced by ALTO. This will contain a number of frames with different slots. Each frame either stands for a single object in the domain, or a class of such objects. Although CommonSloop makes a distinction between classes and instances, this distinction is not made in ALTO, where each object is treated as a class. The transformation algorithm will first establish for each leaf-node in the object hierarchy stored in the CommonSloop knowledge base, whether it is an individual object, or a class of objects. However, this cannot be done automatically and we must therefore ask the user for each leaf node what the intended interpretation is. Individual objects are then translated as constants in the CKRL, whereas classes are translated as one-place predicates. Moreover, class leaf nodes are assumed to be non-empty, and we therefore add an axiom saying that there exists some object with the property denoted by the class-frame.

Once the translation of each frame is determined, other parts of the translation can be done automatically. For example, a super-link between a frame corresponding to a constant a and a frame corresponding to a predicate p is simply translated as $(p(a))$, whereas a super-link between two "predicate" frames p and q is translated as $((\supset x)(p(x) \rightarrow q(x)))$.

Slots can normally be translated automatically; a slot corresponds to a two-place predicate. The exact translation of a slot and its filler depend on the type of filler. If the filler is a simple value, then translation is straightforward. A slot s with value v associated with a "constant" frame a is simply translated as $(s(a,v))$, whereas a slot s with value v associated with a "predicate" frame p translates as $((\supset x)(p(x) \rightarrow s(x,v)))$. However, if the value of a slot is a pointer to another frame, then the translation needs to be more complex. If the value of the slot points to a "constant" frame, then we can simply use the name of this frame. This is the case in formula 1 below where we assume that the value of the *treated-by* slot associated with the frame p is a pointer to the constant frame k. However, the situation is more complicated if the value of the slots points to a "predicate" frame. In this case, interaction with the user is necessary to determine the intended interpretation. Typically, the intended interpretation will be that for each instance i of the frame the slot is associated with there is an instance v of the frame the value points to, such that i stands in the relation denoted by the slot to v. We call this the "existential" interpretation. However, in some cases the intended interpretation is "universal": each i stands in the relation to each v. Thus, assume that the value of the treated slot associated with the frame p is a pointer to the predicate frame drug. Formula 2 illustrates the existential interpretation while 3 illustrates the universal interpretation.

1. $((\supset x)(p(x) \rightarrow treated(x,k)))$
2. $((\supset x)(p(x) \rightarrow (\exists y)(drug(y) \wedge treated(x,y))))$
3. $((\supset x)(p(x) \rightarrow (\supseteq y)(drug(y) \rightarrow treated(x,y)))$

A further complication in the translation of slots is necessary because CommonSloop supports default inheritance. It may happen that the value of a slot is overridden lower down in the hierarchy. A simple translation algorithm might therefore generate an inconsistent logical knowledge base, even though the input CommonSloop knowledge base was consistent under the intended interpretation. The translation algorithm takes care to avoid this by first determining whether any of the descendants of the frame have an incompatible slot value. If so, these are explicitly stored as exceptions in the antecedent of the rule.

In addition to generating axioms from super-links and slots, the Laddered Grid translation algorithm also uses the sub-links to generate further axioms. However, whereas super-link and slot axioms could to a large extent be generated automatically, the sub-link axioms require extensive interaction with the user. Two types of axioms are attempted. First, the translation algorithm will ask the user if the children of some frame form an exhaustive list. If the answer is affirmative, it adds the axiom to the effect that any entity that is an instance of the parent frame must be an instance of at least one of the children as well. Second, the translation algorithm will try to establish whether the children of a parent are disjoint. In a number of cases, the algorithm can establish for itself that this is not the case, for example when two children have descendants in common (Note 4). If the translation algorithm cannot determine the disjointness or otherwise of two frames, it will ask the user. If the user affirms disjointness, the translation algorithm adds an axiom to this effect. In both cases, nothing is done if the user answers negatively.

Fig. 5.8 contains a simple CommonSloop knowledge base, whereas Fig. 5.9 contains the corresponding translation into logic. We have assumed that the user has decided that all leaf nodes are classes. Moreover, as can be seen from axioms 10 to 13, the user has answered affirmatively to all questions concerning exhaustiveness and disjointness asked by the system (Note 5).

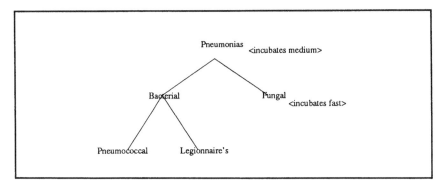

Fig. 5.8. A simple CommonSloop knowledge base.

The translation from the card sort tool into logic is a subset of the translation from CommonSloop into logic: each card corresponds to either an individual in the domain or a class of such individuals, and they are translated accordingly as constants or 1-place predicates. Each sortal criterion can be seen as an additional 2-place predicate over the set of entities. The names of the different categories correspond to the values of the different objects for this predicate. Thus, sorting a set of cards denoting diseases in terms of *severity,* with sorts *minor, average* and *major* for example, gives an additional 2-place predicates whose

1. $((\exists x)(fungal(x)))$
2. $((\exists x)(pneumococcal(x)))$
3. $((\exists x)(legionnaire's(x)))$
4. $((\forall x)(fungal(x) \rightarrow pneumonia(x)))$
5. $((\forall x)(bacterial(x) \rightarrow pneumonia(x)))$
6. $((\forall x)(pneumococcal(x) \rightarrow bacterial(x)))$
7. $((\forall x)(legionnaire's(x) \rightarrow bacterial(x)))$
8. $((\forall x)((pneumonia(x)\&\neg fungal(x)) \rightarrow incubate(x, medium)))$
9. $((\forall x)(fungal(x) \rightarrow incubate(x, fast)))$
10. $((\forall x)(pneumonia(x) \rightarrow (fungal(x) \vee bacterial(x))))$
11. $((\forall x)(fungal(x) \rightarrow \neg bacterial(x)))$
12. $((\forall x)(bacterial(x) \rightarrow (pneumococcal(x) \vee legionnaire's(x))))$
13. $((\forall x)(pneumococcal(x) \rightarrow \neg legionnaire's(x)))$

Fig. 5.9. The translation of the simple CommonSloop knowledge base.

first argument describes a disease and whose second argument is minor, average or major. The exact translation of sorting a card into a category depends on whether the card corresponds to an instance or a class of entities in exactly the same way as in CommonSloop. In addition, the translation algorithm will query the user whether the different sorts that have been given are in fact exhaustive. Since the card-sort tool allows the user to use a "dunno" category (which is obviously not translated), the translation algorithm knows for each sort that contains this category, the sort is not exhaustive. For other sorts, the user is queried.

The translation from the repertory grid into logic is relatively straightforward as well. Each element is again translated as a constant or as a predicate. Each construct is a two-place predicate whose first argument corresponds to the element in the grid and whose second argument corresponds to the value of the element on this construct. The results of entailment analysis establish that a range of values on one construct implies a range of values on another. We translate a range of values in the obvious way as a disjunction. The entailments that result from the entailment analysis are then translated as logical implications. Thus, if the entailment analysis shows that a value in the range [1,2] on the severity construct entails a value in the range [4,5] on the incubation construct, this would be translated as $((\supset x)((severity(x,1) \vee severity(x,2)) \rightarrow (incubation(x,4) \vee incubation(x,5))))$.

The current translation algorithm for the repertory grid tool has the disadvantage that it introduces numerical values. The problem is that the other tools tend to use symbolic values. This therefore severely restricts the synergy between the repertory grid tool and the other tools in ProtoKEW. One way of overcoming this problem would be to make the translation algorithm for the repertory grid more sophisticated. One could imagine a translation algorithm that queried the user in order to find an explicit semantic translation for the rating on constructs. Thus, in the above example, the more complicated translation algorithm would query the user in order to determine a symbolic term for values in the range [1,2], for example *minor*.

AQ11 output can be translated into logic in a fully automatic way. It is stored in the form of a decision tree whose leaf nodes correspond to the different classes. Each leaf node is therefore translated as a 1-place predicate. The branches coming out of a node correspond either to the attribute or to the negation of the attribute. An attribute is translated as a 1-place predicate. The decision tree can then be translated by translating paths from the root node to each leaf node as an implication where the consequent of the translation of the leaf node, and the antecedent is a conjunction. Each conjunct corresponds to the negation of an attribute, in which case the corresponding predicate is either negated, or the presence of an attribute, in which case the predicate occurs positively.

The various translation algorithms have a number of features in common. In most cases, they have to be interactive. In general, tool-specific knowledge representation languages may leave the exact interpretation of some feature implicit, and in order to make these explicit one may need to consult the user. Moreover, most translation algorithms are straightforward to write once the interpretation of the basic entities in the tool-specific knowledge base and its use have been made explicit.

8. INTEGRATION

In the previous section, we described the way in which ProtoKEW transforms knowledge formulated in a tool-specific knowledge base into logic, the language used by the Common Knowledge Base. However, simply transforming knowledge is not enough. ProtoKEW also needs to integrate knowledge from different sources. In this section, we discuss the problems of integration in more detail.

As noted in section 5, the Common Knowledge Representation Language supplies a partitioning mechanism. This partitioning mechanism is used to store transformed tool-specific knowledge bases in the CKB; each such knowledge base is stored in a separate partition. Since the CKRL interpreter can be instructed to prove a proposition using information in more than one partition, or to forward chain on more than one partition, knowledge can be integrated in this simple way.

However, the danger of this simple form of integration is of course that even though the information in each individual partition may be consistent, the combined information in several partitions may be inconsistent. Obviously, trying to prove things from inconsistent knowledge base is undesirable. ProtoKEW therefore supports tighter integration by merging two partitions into a single partition. Merging amounts to copying the contents of one partition to a new partition and then incrementally adding propositions from the other partition. While adding these new propositions, both tests discussed in section 5 are performed. Thus, we ensure that the new propositions are not redundant, and cannot be derived from the information already stored in the partition, and that the new propositions are consistent with the existing established propositions.

If the theorem prover discovers that a new proposition is not consistent with the propositions in a partition, the reason maintenance system is called. We have slightly extended the RMS. Usually, a justification stored with a proposition is simply a list of propositions, which together allow one to prove the proposition in question, or an empty list, if the proposition is a premise. In ProtoKEW's CKRL, we also maintain pointers to the partition from which the proposition originated. Because in general partitions will be created as the result of transforming a tool-specific knowledge base, this helps trace a piece of knowledge back to its source.

The information concerning the origin of a particular piece of information can be used by the RMS. In most RMSs, if the system discovers a contradiction, control is passed to the user who will then have to decide which of the contradictory propositions to withdraw. In ProtoKEW's RMS, if one source is believed to be more reliable than another, propositions originating from this source will always be retained in preference to those from a less reliable source.

9. EVALUATION

One of the major factors the modeller uses in their selection of the appropriate knowledge acquisition tool is their assessment of the *quality* of the knowledge in the CKB. A requirement on ProtoKEW must be the evaluation of the partitioned and integrated knowledge bases. Evaluation is necessary because the knowledge base is constructed with some particular aim in mind, and the knowledge acquisition process can be terminated if the knowledge base allows one to achieve this aim. In the context of expert systems, the aim of a knowledge base is typically two-fold. First, the expert system must be able to solve particular problems. Second, the expert system must be able to generate satisfactory explanations and justifications of its solutions.

Evaluation in ProtoKEW's CKB is restricted to a KB's problem solving behaviour. This is in accordance with most of the literature. One can distinguish between two types of criteria that a satisfactory knowledge base must meet for problem solving purposes. First, there are a number of internal criteria, such as consistency and non-redundancy. Second, there is the question of veracity of a knowledge base; a knowledge base can in principle be both consistent and non-redundant without giving satisfactory answers, simply because some of the knowledge in it is false, or because some piece of knowledge is missing. The second test has also been called "validation of the knowledge base".

Internal criteria of consistency and redundancy pose no problems forProtoKEW. In fact, one of the reasons for choosing logic as the CKRL was to ensure that it would be relatively straightforward to assure that knowledge bases would meet these internal criteria. Logic has two advantages in this respect. First, it provides one with a clear and non-ambiguous definition of notions like consistency and redundancy which is independent of the interpreter used for inference. This is not straightforward for production rule based systems. For example, is the rule set below inconsistent? We cannot answer this in a production system without reference to the conflict resolution methods used by the interpreter. If one uses specificity as the conflict resolution strategy, then the rules do not necessarily lead to inconsistencies; if one uses an exhaustive forward chaining control regime, then the rules can lead to inconsistencies (Note 6):

1. IF ϕ THEN add(ψ)
2. IF ϕ AND ϕ' THEN add ($\neg \psi$)

Another complication is that in a production rule system, information is stored both in the rule base and in working memory. However, only the rule base is permanent, and certain inconsistencies can be detectable only if some information has been stored in working memory. Thus,

IF ϕ THEN add(ψ)
IF χ THEN add($\neg \psi$)

is inconsistent only if working memory contains both ϕ and χ. However, such information is usually not available when a rule base is constructed. One form of consistency checking therefore consists of generating descriptions of all possible states of working memory and asking the expert whether such states are possible (Rousset, 1988). In ProtoKEW this problem does not arise because all information is stored in the same format, and all incoming information first undergoes a redundancy and consistency check (section 5).

A second advantage of logic is that one can in principle use the same interpreter both for performing consistency and redundancy checks as well as for problem solving; ProtoKEW does in fact use the same interpreter for both. This is in sharp contrast with Ginsberg (1988) and Nguyen, Perkins, Laffey and Pecora (1985) who propose separate mechanisms for consistency and redundancy checking on the one hand, and problem solving on the other.

The second type of evaluation that one wants to perform on a knowledge base is validation: is the information stored in the knowledge base true? One obvious way of validating a knowledge base might be to present the different propositions in the knowledge base to the expert. While this may in general allow one to detect certain false propositions, it cannot guarantee that we find all errors. Moreover, we are not guaranteed to find missing pieces of knowledge. An alternative is to present the knowledge base with a number of test cases, and to observe whether it provides the correct solutions for these tests.

If one adopts the second form of validation, then two further problems arise. First, one needs to generate a set of test cases which is as complete as possible. Second, given a test case and an incorrect solution to this problem, the knowledge base needs to be refined.

The problem of generating a satisfactory set of test cases can itself be seen as a knowledge acquisition process. There are a number of systems around that analyze a knowledge base (usually represented as a set of production rules) to generate a set of test cases. For example, Rousset (1988) and Vignollet and Ayel (1990) generate sets of test cases by automatically generating all possible states of working memory which could lead to some terminal fact (i.e. some fact occurring only on the left hand-sides of a set of production rules). These states of working memory, together with the terminal fact, can then be presented to the expert for validation. Unfortunately, this approach is not feasible in ProtoKEW. ProtoKEW's CKRL is full first-order logic. As a result, the notion of a terminal fact, which is crucial to both Rousset's and Vignollet and Ayel's work, cannot be defined for ProtoKEW. An implication like $\phi \rightarrow \psi$ can be used both to prove ψ from ϕ and $\neg \phi$ from $\neg \psi$. Another shortcoming of the method is that, although we could in principle detect the conditions under which a conclusion could be falsely derived, we cannot detect incompleteness in the sense that a conclusion could not be derived, even though it should be true. It is on the basis of these considerations that we argue that the automatic generation of test cases for a system with the expressive power of ProtoKEW is not feasible. ProtoKEW therefore provides no direct support for this.

Generating a set of test cases is only one half of the problem of validation. Once a set of test cases has been generated, we need to develop tools for supporting refinement of the knowledge base. Thus, if the system is able to derive a false conclusion, tools must be provided for retracting whatever propositions were responsible for the conclusion, and, conversely, if the system fails to derive a true conclusion, tools must be provided to add the relevant knowledge. ProtoKEW provides some very limited support for these two activities. When asked to derive a proposition, ProtoKEW will display the proposition that were used in the derivation. If the proposition to be derived is false, the user is then able to decide which of the used propositions was at fault, and retract it. In the retraction of propositions, we make use of the RMS capabilities of the CKB in order to ensure that consistency is maintained.

The support for the second activity (adding further propositions) is more rudimentary. The CKB has a tracing mechanism that allows one to observe the behaviour of the theorem prover during the search of a proof. Clearly, when a proof fails, inspecting this trace will give one some indication as to where the incompleteness in the CKB arose.

A number of more sophisticated refinement tools are described in the literature. For example, both Craw and Sleeman (1990) and Aben and van Someren (1990) describe refinement programs which suggest changes to a backward chaining production rule system in order to make the system's conclusions agree with those elicited from an expert. Both systems, as implemented, assume that the knowledge is stored as a set of backward chaining production rules and that textual order is used for conflict resolution. Since our CKRL is more powerful, we would need to generalize the methods used by Craw and Sleeman and Aben and van Someren. In our current work we are investigating the possibilities of doing so.

Before leaving the topic of evaluation we should note that our discussion has focused on evaluation of CKB partitions. In fact, another strength of ProtoKEW is that we allow our individual knowledge acquisition tools to conduct their own analysis and evaluation of the knowledge that is accruing from a particular acquisition session. This evaluation exploits features of the tool's own representational language. In ALTO, for example, a number of tests can be performed on the knowledge being accrued, such as tests for discrimination between classes or instances, tests for the generality of attributes and so on (Note 7). These tests are facilitated by the representation used by the particular knowledge acquisition tool. The structured objects of ALTO's CommonSloop representation make it much easier to reason about, for example, discrimination between classes. This type of test is cumbersome in our CKRL because the properties of objects become dispersed throughout the logical sets of propositions describing the results of laddering.

Our experience points clearly to the fact that certain types of knowledge are best evaluated in the context of the tools that elicit them, whilst other sorts of evaluation are more properly conceived of as tests on sets of integrated theories describing the application domain.

10. LIMITATIONS AND FUTURE WORK

In this paper we have described work underway to construct an integrated knowledge acquisition workbench. The ESPRIT II ACKnowledge Project (P2572) has provided the stimulus for this work, and will in turn, integrate the results from a range of such prototype experiments, and produce a substantial software deliverable (KEW).

In terms of recent experiments at Nottingham we have gained some insight to the ways of building integrated knowledge acquisition software support tools. We have attempted to understand how the knowledge acquisition process can be informed and directed by knowledge about knowledge acquisition tools, the knowledge acquisition process, models of problem solving, tasks, and domains. We believe that central to this endeavour is the construction of workbenches founded on a clear understanding of the semantics of their various components.

In future work we are looking to incorporate additional knowledge acquisition tools, and to extend the ways in which existing tools can be used. The aim is to produce configurable knowledge acquisition tools that can be adapted in a number of ways. Many of the tools in ProtoKEW embody oversimple assumptions about how they might be used. For example, ALTO assumes that one always ladders on objects and that the only structural relation between objects are subset and member-of links. However, one often uses manual forms of the laddering technique to elicit part-of hierarchies. Alternatively, one may ladder on attributes of objects.

Building configurable tools will in turn necessitate a wider range of translation methods from our tools into the logic that is the core of our workbench. For example, if one ladders on the attributes of objects, the structural relation might be something like "is implied by". If one uses a laddering tool to elicit a part-of hierarchy, then the structural relation would be translated by means of the predicate part-of. In either case, the translation algorithm would need to be changed. It is fairly clear what these changes would have to be. For example, in terms of the ALTO translation algorithm in ProtoKEW, this would mean allowing (i) a range of different possible translations of nodes, (ii) a number of different possible translations of the structural links between nodes.

Whilst for laddering it might be possible to foresee the different ways in which a knowledge engineer might want to use this tool, some knowledge acquisition tools allow for a much wider range of less routined and more creative uses. For example, although one will typically use a repertory grid with objects or classes of objects as elements, one could in principle allow the elements to be more complex. One could for instance use as elements dependencies between objects. In this case, the repertory grid would allow one to elicit knowledge about properties of dependencies between objects, that is, relations between objects.

Nevertheless, once one knows the interpretation of the basic elements in the grid, it is relatively straightforward to see how to translate the information one gets out of the grid. However, unlike the laddering tool, it is less clear that one can specify the possible uses of the repertory grid in advance. The consequence is that a knowledge engineer who wants to use a particular tool in a novel fashion has to be prepared to change the translation algorithm accordingly, or do the translation manually.

We are also interested in increasing the amount of synergy between tools. For example, the use of rating techniques in conjunction with machine learning methods may give us a way of preprocessing training sets for machine learning tools so that the training examples are organised more knowledgeably. This might lead to the suppression of some attributes in the case descriptions in favour of other more relevant features, where the notion of relevance is determined by the expert as the result of using a rating tool.

Another obvious area of development is an increased range and use of directive models. In particular, we are considering the use of interpretation models (abstract models of the problem solving architecture) as a means of organising the partitions in a CKB. Thus partitions might be constructed in terms of the problem solving role that the knowledge plays.

Finally, we are looking to develop the Knowledge Engineering Knowledge Base into a more active component of ProtoKEW. Currently, it is mainly used as a repository for the compiled knowledge structures used in the directive models and knowledge acquisition goal trees. We hope to provide a more dynamic aspect to the KEKB. That is, we hope to use knowledge about the knowledge acquisition process that actively monitors what the user is doing, how the CKB partitions are evolving, and so on, to enable the KEKB is able to offer advice on the courses of action that the knowledge engineer might want to consider. Such advice heralds the emergence of real knowledge-based systems for knowledge acquisition.

REFERENCES

Aben, M. & van Someren, M. (1990) Heuristic refinement of logic programs. ECAI-90, 7-12.

Anjewierden, A. (1987) Knowledge Acquisition Tools. AI Communications, Vol 0, No 1.

Anjewierden, A. & Wielemaker, J. (1989) An Architecture for Portable Programming Environments. Proceedings NACLP'89 Workshop on Logic Programming Environments, 10–16.

Boose, J. (1985) A knowledge acquisition program for expert systems based on personal construct psychology. Int. J. Man-Machine Studies 23:495–525.

Boose, J. (1986) Rapid acquisition and combination of knowledge from multiple experts in the same domain. Future Generation Computing Systems 1:191–216.

Breuker. J. &. Wielinga, B. (1987) Use of models in the interpretation of verbal data. In A.L. Kidd, (ed), Knowledge Acquisition for Expert Systems, a practical handbook, New York: Plenum Press.

Breuker. J. &. Wielinga, B. (1989) Model driven knowledge acquisition. In P. Guida and G. Tasso, (eds), Topics in the design of expert systems, Amsterdam:. North Holland.

Burton, A., Shadbolt, N., Hedgecock, A. & Rugg, G. (1987) A formal evaluation of knowledge elicitation techniques for expert systems. In D Moralee, (ed), Research and development in expert systems, IV, 136–145.

Burton, A., Shadbolt, N., Rugg, G. & Hedgecock, A. (1988) Knowledge elicitation techniques in classification domains. ECAI-88, 85–90

Bylander, T. & Chandrasekaran, B. (1988) Generic tasks in knowledge-based reasoning: The "right" level of abstraction for knowledge acquisition. In B. Gaines and J.Boose, (eds), Knowledge Acquisition for Knowledge Based Systems, volume 1, pages 65–77. Academic Press, London, 1988.

Clancey, W. (1983) The epistemology of a rule based system -a framework for explanation. Artificial Intelligence 20:215–251.

Clancey, W. (1985) Heuristic classification. Artificial Intelligence, 27:289–350.

Craw, S. & Sleeman, D. (1990) Automating the refinement of knowledge-based systems. ECAI-90, 167-172.

Elfrink, B. & Reichgelt, H. (1989) Assertion-time inference in logic-based systems. In P. Jackson, H. Reichgelt & F. van Harmelen, (eds), Logic-based Knowledge Representation. Boston: MIT Press, Boston.

Eshelman, L. (1989) MOLE: A knowledge-acquisition tool for cover-and-differentiate systems. In S. Marcus, (ed), Automating Knowledge Acquisition for Expert Systems, pages 37–80. Dordrecht: Kluwer Academic Publishers.

Friedland, P. (1979) Knowledge-based experiment design in molecular genetics. Report STAN-CS-79-771. Stanford University.

Gaines, B. R. & Shaw, M. L. G. (1986) Induction of inference rules for expert systems. Fuzzy Sets and Systems, 8 (3), 315–328.

Gaines, B. & Linster, M. (1990) Development of second generation knowledge acquisition systems. In B. Wielinga, J. Boose, B. Gaines, G. Schreiber & M. van Someren, (ed),Current Trends in Knowledge Acquisition, pages 143–160. Amsterdam: IOS Press.

Ginsberg, A. (1988) Knowledge-base reduction: A new approach to checking knowledge bases for inconsistency and redundancy. AAAI-88, 585-589.

Hayes, P. (1977) In defence of logic. IJCAI-5. 559–565.

Jackson, P., Reichgelt, H. & van Harmelen, F. (eds), (1989) Logic-based Knowledge Representation. Boston: MIT Press, Boston.

Kelly, G. (1955) The psychology of personal constructs. New York: Norton.

Major, N. & Reichgelt, H. (1990) Alto: An automated laddering tool. In B. Wielinga, J. Boose, B. Gaines, G. Schreiber & M. van Someren, (ed),Current Trends in Knowledge Acquisition, pages 222–236. Amsterdam: IOS Press.

Marcus, S. (1988) Automatic knowledge acquisition for expert systems. New York: Kluwer.

McAllester, D. (1980) An outlook on truth maintenance. Technical report, MIT AI LAB.

Michalski, R. & Larson, J. (1978) Selection of most representative training examples and incremental generation of VL1 hypotheses: The underlying methodology and the description of programs ESEL and AQ11. Rep. No. 867. Computer Science Dept., University of Illinois. Urbana.

Mitchell, T. (1979) An analysis of generalization as a search problem. IJCAI-6, 577–82.

Moore, R. (1984) The role of logic in Artificial Intelligence. SRI AI Center, Technical Note 335.

Motta, E., Rajan, T., Domingue, J. & Eisenstadt, M. (1990) Methodological foundations of KEATS, the knowledge engineer's assistant. In B. Wielinga, J. Boose, B. Gaines, G. Schreiber & M. van Someren, (ed),Current Trends in Knowledge Acquisition, pages 257–275. Amsterdam: IOS Press.

Newell, A. (1982) The knowledge level. Artificial Intelligence, 18:87–127.

Nguyen, T., Perkins, W., Laffey, T. & Pecora, D. (1985) Checking an expert system's knowledge base for consistency and completeness. IJCAI-9, 375-378.

Reichgelt, H., Major, N. & Jackson, P. (1990) Commonsloop: The manual. Technical report, AI Group, Dept Psychology, University of Nottingham.

Rousset, M.-C. (1988) On the consistency of knowledge bases: The COVADIS system. ECAI-88, 79-84.

Shadbolt, N. & Burton, M. (1989) The empirical study of knowledge elicitation techniques. SIGART Newsletter, 108, April 1989, ACM Press.

Shadbolt, N. & Burton, M. (1990) Knowledge elicitation. In J. Wilson and N. Corlett, (eds), Evaluation of Human Work: A Practical Ergonomics Methodology, pages 321–346. Taylor and Francis.

Shadbolt, N. & Wielinga, B. (1990) Knowledge based knowledge acquisition: the next generation of support tools. In B. Wielinga, J. Boose, B. Gaines, G. Schreiber & M. van Someren, (ed), Current Trends in Knowledge Acquisition, pages 313–338. Amsterdam: IOS Press.

Steels, L. (1990) Components of expertise. The AI Magazine, 11:30–62.

Vignollet, L. & Ayel, M. (1990) A conceptual model for building set of test samples for knowledge bases. ECAI-90, 667-672.

ACKNOWLEDGEMENTS

We would like to thank Nigel Major for his help in implementing aspects of ProtoKEW, to Peter Terpstra of the University of Amsterdam and GEC Marconi Research Centre who made available software for inclusion in ProtoKEW. Thanks are also due to Derek Sleeman who commented on an earlier draft of this paper. This research was carried out, in part, under the auspices of ESPRIT P2576 ACknowledge.

NOTES

1. The ACKnowledge consortium comprises: Cap Sesa Innovation, Marconi Command and Control Systems, GEC-Marconi Research Centre, Telefonica, Computas Expert Systems, Veritas Research, the University of Amsterdam, Sintef, and the University of Nottingham.

2. In Shadbolt and Wielinga (1990) we describe how this selection might ultimately be automated.

3. It is in general impossible to determine automatically whether a set of propositions in classical first-order predicate calculus is consistent. Any theorem prover needs to contain heuristics to halt the search for a proof. Because these heuristics may terminate the search for a proof too early we can never guarantee consistency for any set of propositions in KB.

4. This can occur because CommonSloop supports multiple inheritance.

5. We do not want to suggest that this KB is veridical.

6. It is interesting to note that Ginsberg's (1988) technique for checking rule-based knowledge bases can only be applied if we assume that the interpreter uses an exhaustive forward-chaining control regime.

7. Details of the various types of evaluation specific to the results of laddering can be found in Major and Reichgelt (1990).

CHAPTER 6

POPLOG's Two-level Virtual Machine Support for Interactive Languages

Robert Smith, Aaron Sloman*, John Gibson
School of Cognitive and Computing Sciences,
University of Sussex, Brighton, England
*Now at School of Computer Science, the University of
Birmingham, England.

ABSTRACT

Poplog is a portable interactive AI development environment available on a range of operating systems and machines. It includes incremental compilers for Common Lisp, Pop-11, Prolog and Standard ML, along with tools for adding new incremental compilers. All the languages share a common development environment and data structures can be shared between programs written in the different languages. The power and portability of Poplog depend on its two virtual machines, a high level virtual machine (PVM—the Poplog Virtual Machine) serving as a target for compilers for interactive languages and a low level virtual machine (PIM—the Poplog Implementation Machine) as a base for translation to machine code. A machine-independent and language-independent code generator translates from the PVM to the PIM, enormously simplifying both the task of producing a new compiler and porting to new machines.

1. INTRODUCTION

During the late 1970s and early 1980s the AI group at Sussex University needed an AI environment for teaching and research. We adopted the policy of using general-purpose computers (initially PDP11, then VAX (first running VMS, then later Unix), then Sun and other workstations) rather than Lisp machines partly because we could not possibly afford enough Lisp workstations for all our staff and students, and partly because we desired maximum flexibility and independence from particular manufacturers.

The system had to be approachable enough for totally naive first-year students, including non-numerate Arts and Social Studies students, yet powerful enough to support advanced teaching and our own research and software development. It had to be usable both on time-shared machines with dumb VDU interfaces, and on powerful workstations with bit-mapped displays, including those supporting the X Windows system. Both our research and teaching required access to a variety of AI languages, and some of our projects required mixed language programming (e.g. Pop-11 and Prolog) so the system had to support several different languages sharing a common development interface, common data-structures, etc. Because so many AI projects do not start from a well-defined specification, we wanted a system that supported rapid prototyping and exploratory development for the purpose of clarifying a problem, which meant using interpreters or incremental compilers, preferably with the convenient interface of an integrated editor. Because we were a large and growing AI community facilities for sharing and re-using software and documentation were important.

An interface to subroutines written in conventional languages such as C, Fortran or Pascal, or libraries such as NAG, was also required especially for work in speech or vision.

Moreover, because AI research continually points to the need to develop new languages tailored to specific applications, the system had to make it easy to implement new languages with a good development environment.

And finally, because new machines were becoming available it had to be comparatively easy to port.

This demanding array of requirements is, as far as we know, not met by any other system. Some operating systems, like VMS and Unix provide mixed language development environments, but they directly support only "batch-compiled" languages, not incremental compilation. That is to say, program files have to be compiled to object files, then later all the object files are linked into a runnable image. By contrast an incremental compiler allows the source language to be used as a command language and allows procedures to be edited and re-compiled or new procedures added without re-linking the whole system. This enormously speeds up development and testing of software.

Unix does have the advantage of being portable, but anyone developing a new compiler to run under Unix will normally have to write the code-generator for each new host machine. This is not required if, for instance, an interpreter is written using a language like C, which is normally provided with Unix systems, but for many purposes an interpreted language will run too slowly, compared with a compiled version.

Several AI systems based on Lisp have provided more than one language, in an interactive development system, but this is usually done by writing, in Lisp, interpreters for the other languages. Although programs written in Lisp can run fast in such a system (provided that the Lisp system is well engineered) programs written in the other languages will be slowed down significantly on account of being interpreted. By contrast the Poplog architecture does not favour one language over others. All the languages are compiled to machine code for maximum speed of execution (though of course interpreters can be written if required, e.g. for reduced space or increased flexibility).

Poplog allows data-structures to be shared between programs written in different languages because they run in the same process, with the same address space. For some kinds of programs, especially where there is no requirement for very rapid communication between sub-systems, it might be preferable to have different languages running in different processes, which would allow distribution over different machines.

The remainder of this paper describes the two-level virtual machine architecture of Poplog, which makes it possible to meet all the requirements sketched above.

2. THE POPLOG VIRTUAL MACHINES

Like other compiler systems that must accommodate several languages and be portable across a wide range of computer architectures, Poplog employs an intermediate virtual machine. Steel [1960] proposed the use of an UNCOL (UNiversal Computer Oriented Language) to reduce the effort required to implement a new language or to produce code for a new architecture. Using this, to implement L languages on M machines requires L front ends (each translating one source language to the intermediate language) and M back ends (each translating the intermediate language to one machine language) rather than having to write L*M complete compilers.

The implementors of the Amsterdam Compiler Kit (ACK), Tanenbaum et al. [1983], claim that the original UNCOL concept failed because of a desire to accept all languages and all machine architectures using a single virtual machine language. The success of the ACK, which uses a compiler intermediate virtual machine, EM, is claimed to be due to restricting the system to algebraic languages and byte-addressable machines. Poplog is currently only implemented on byte-addressable machines with a 32-bit word size. This restriction could be relaxed (although there may be some efficiency penalties as discussed in section

C.2.). So far, however, no processor without these attributes has appeared for which there was a need to port Poplog.

The range of languages supported by Poplog is much more diverse than that of the ACK. These are Pop-11 (a Lisp-like language with a Pascal-like syntax), Prolog, Common Lisp, Standard ML and sysPOP (an extended dialect of Pop-11 used for the Poplog system sources, as described later). Users have implemented additional languages and tools, such as Scheme (a statically scoped dialect of Lisp), KDL (a frame-based knowledge description language), Flavours (an Object-Oriented extension to Pop-11), KERIS (a collection of knowledge engineering tools extending Common Lisp), RBFS (a Rule Based Frame System produced by Brighton Polytechnic), a vision toolkit developed at Reading University, and various special purpose languages, e.g. a control language developed for a real time expert system project. FLEX, a Prolog-based expert system toolkit developed by Logic Programming Associates, is now available on Poplog, as is RULES, a rule induction program produced by Integral Solutions Ltd.

One apparent disadvantage of Poplog at present is that from the point of view of users it does not support languages, like C, that permit direct pointer manipulation, since the automatic garbage collection mechanism could at any time re-locate an object thereby invalidating a pointer to one of its fields. This restriction is removed in sysPOP, which allows knowledgeable system programmers to manipulate pointers and offsets, since these mechanisms are required for implementing Poplog in any case. It would also be possible to add a pointer data-type to enable user programs to simulate pointer manipulation. This would make it possible, for instance, to add an incremental compiler for a language like C to Poplog, in order to speed up development of C programs.

An improvement to the original UNCOL strategy is the use in Poplog of two virtual machines rather than one. The connection between these is shown in Fig. 6.1.

All the compiler front ends produce code for the same high level Poplog Virtual Machine (PVM). The high level VM is strongly language oriented and consequently the front ends are straightforward to write. The high level VM code is translated to code for the low level Poplog Implementation Machine (PIM). The PIM is machine oriented, and translating from its virtual machine instructions to target machine code is trivial: indeed many only require one instruction on the VAX.

As will be explained later, there are actually two different compilation routes between the PVM and the PIM, one used for incremental compilation of user procedures, and one used for batch compilation of the system sources, which are mostly written in the sysPOP dialect of Pop-11. Much effort can be spent on improving the phase of code

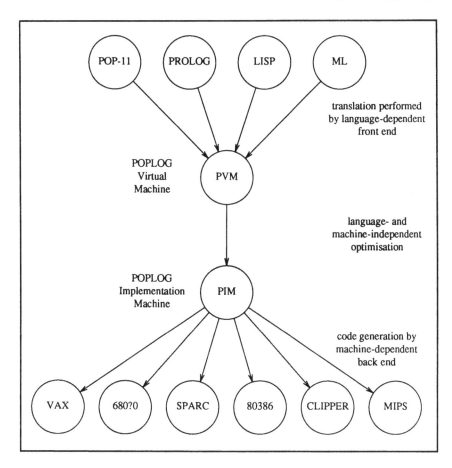

Fig. 6.1.   The Poplog Virtual Machines.

generation between PVM and PIM because this translation is both language and machine independent, so work done at this level benefits all implementations.

The next two sections describe the data areas in Poplog and the scheme for data representation. This will help in understanding the more detailed description of the two virtual machines which then follow.

2.1   Object Representation

Most of the Poplog languages are untyped and thus type-checking must occur at run time. The following scheme for data object representation is used for all current implementations of Poplog, these being on 32-bit

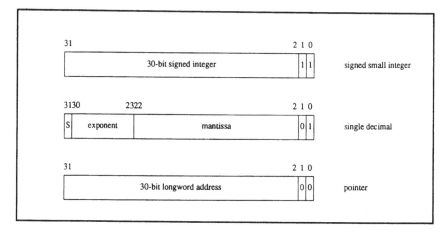

Fig. 6.2.   The three forms of tagged object reference in Poplog.

machines with byte addressing. Each reference to a Poplog object is held as a 32-bit quantity. There are three forms, as shown in Fig. 6.2, which are differentiated by a 2-bit tag.

This shows that small signed integers and single decimals are held as immediate values in the object reference. Other objects are stored as longword-aligned vectors of longwords in memory, which are addressed by the pointer in the object reference. The type of these objects is determined by the contents of the second field of the vector, the "key" field, which is actually itself a pointer to a structure that defines the type of the object. If two objects have a matching key field then they are of the same type. The object reference actually points to the third longword of the object. This equates to the start of data for strings and other vector-based types, thus allowing these to be passed unmodified to and from "externally loaded" (i.e. imported) foreign procedures written, for example, in C or Fortran.

Not all of the object vector need contain other object references. Signed and unsigned bitfields, bytes, words, longwords and machine instructions (in the case of procedure objects, which are first class objects) can be present, and information about this structure, held in the key, will be used by the garbage collector when tracing live objects. As an example, Fig. 6.3 shows the representation of the two element list ['foo', 1].

Each key structure contains information about the class of objects which it defines, including functions defined for that data type: e.g. constructor, accessor and print functions. A large number of data types are provided, and the user is free to create new compound types of vectors or records, and to change some of the class-specific functions,

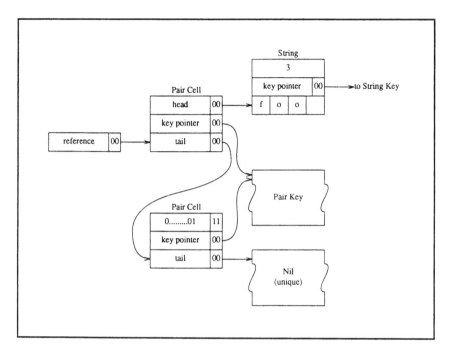

Fig. 6.3.   Schematic diagram showing the representation of the two element list.

e.g. functions for printing, equality testing, or the "class_apply" function which determines what is to happen if a structure is treated as a procedure and executed.

The advantages of the current scheme of object representation are:

• A pointer to an object (including the tag bits) is equivalent to the address of the third longword of that object, and thus no tag manipulation is required in this case. Additionally, only the least significant bit of the reference need be tested to determine if a reference is a pointer or a simple object.

• Most architectures can test, insert and extract the 2 least significant bits efficiently with quick arithmetic/logical instructions. Other tagging schemes generally require shift operations to implement, which are generally slower.

• The 2 tag bits of a pointer reference are available to the garbage collector to hold the mark and shift bits during garbage collection.

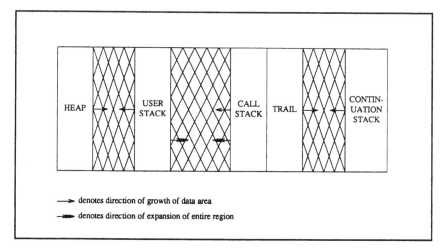

Fig. 6.4.   The Poplog Data Areas.

- The key field of an object can be used to hold the new address of that object during the compaction phase of garbage collection.

Some implementations of AI languages on 32 bit machines have used tags larger than two bits, to enable fast identification of frequently used data-types. This strategy was rejected for Poplog on the grounds that the reduction in available address space would be intolerable for some of the larger potential applications.

2.2   Data Areas and Control Mechanisms

There are five data areas in Poplog, which are contained in two segments of contiguous memory as shown in Fig. 6.4. The base of the heap is positioned after the shared Poplog system code and data, which has not been shown in the diagram. The segment containing the call stack need not grow towards the base of memory as shown, its orientation being determined by that of the call stack.

The "heap" contains all the dynamically allocated objects in the system. The heap is shown as a single entity but in fact it can be segmented, and this allows for the dynamic allocation of memory for "externally loaded" procedures. (For simplicity, the possibility of external segments in the heap is not shown in the diagram.)

Although all live structures in the heap will be scanned during garbage collection, only those above a certain point are eligible for compaction (or removal if dead). The ability to set this point to be the current top of the heap—called "locking" the heap—provides a method of reducing the garbage collection overhead. This is usually done after a large program has been loaded, when all its procedures and data are required to remain live.

In the system data area at the base of the heap is the "dictionary". It performs the role of the symbol table of the conventional batch compiler, which in an interactive system must exist at run time. Unlike strings there must be a unique structure, a "word", for each distinct source language identifier. The dictionary is a hash table which maintains these. Each word record may point to an associated identifier record which defines the run time identifier to which that word (name) is currently bound. The identifier record contains syntactic information about the identifier, and its value, if any. For program modularity, a given word can point to different identifiers in different "sections" of a program.

The "user stack" is used for stacking references to objects. In Pop-11 it is directly accessible from the language, and can be used to build elegant programming solutions. It is used by all languages for passing arguments and results between procedures and for holding temporary values during expression evaluation.

An implementation dependent aspect of maintaining the heap and user stack is that of overflow detection of both entities and underflow detection of the user stack. In operating systems that allow arbitrary memory pages to be made inaccessible, such as VMS and SunOS 4.0, the solution is to create an inaccessible memory page between the two areas and at the base of the user stack. Thus these conditions are detected automatically, and the relevant action is invoked on receipt of the memory violation signal.

Unfortunately this cannot be done with most Unix systems, but aligning the base of the user stack at the end of the data segment makes automatic detection of stack underflow possible. Explicit checks must still be used for overflow detection however, but the overhead of these is not large.

The "call stack" is the conventional call stack of the processor. To allow compatibility with either upward or downward direction of growth of the call stack, Poplog allows the whole segment to be implemented in either direction. However, Poplog enforces its own stack protocol for several reasons:

• Poplog procedures support not only lexically scoped but also dynamic binding of local variables, and the host protocol will not normally cater for this;

• arguments and results are passed on the user stack thus mechanisms for passing these on the call stack are not required;

• and most importantly, various parts of Poplog (e.g. garbage collector, abnormal procedure exit mechanisms) work with stack frames as explicit data structures, and therefore a standard machine independent format is required for these.

Besides the normal call stack discipline supported by most languages Poplog provides a variety of additional mechanisms for manipulating the call stack, without which it would have been impossible to implement Common Lisp or Pop-11.

A commonly used mechanism permits abnormal exit from a procedure, for example if there has been an error interrupt, or if it is required to abort the current procedure and replace it by a call of another, a facility provided by "chain" in Pop-11. User programs can specify that the call stack should be "unwound" up to a specified point and then a new procedure invoked at that point, and so on. A variety of types of condition-handlers can use such mechanisms, including the "catch" and "throw" of Lisp.

It is possible for a procedure to trap such abnormal exits and take action to ensure that necessary tidying up is done, using "dynamic local expressions" as described below. A special case of this is the provision of dynamic local variables in Lisp and Pop-11, which are essentially globally accessible variables whose values are always restored on exit from procedures that declare them as local, a technique that is useful for temporarily altering the interrupt behaviour, or the standard printing channel, or a counter that records the current depth of procedure calls, etc.

In addition, Poplog, like a number of other AI systems, supports a "lightweight" process mechanism. That is to say, it is possible to create a process in which a procedure is run, which may call other procedures and then be suspended. The process record will include information about the state of the call stack and local variables at the time of suspension. Later the process can be resumed, then suspended again, resumed again, and so on. This allows co-routining, and in conjunction with timed interrupts permits a scheduler to control a collection of sub-processes sharing a data-environment — a feature of Poplog that has been used for teaching undergraduates operating system design techniques. It can also be used in AI systems requiring several communicating sub-processes simulating different independent agents or different parts of a single agent.

An unusual, and possibly unique, feature of the Poplog VM is the support for "dynamic local expressions" in conjunction with the above mechanisms for abnormal procedure re-entry and exit. A special case of the mechanism is required for the "unwind protect" facility of Lisp which can trap abnormal procedure exit, but a more general version is available via the "dlocal" construct of Pop-11 in conjunction with the process mechanism as well as abnormal procedure exits.

The "dlocal" construct allows the user to define entry and exit actions to be performed whenever a procedure is entered or re-entered, or exitted. For example, the contents of some datastructure can be stored on entry and restored on exit, or special trace printing can be done to help with debugging of complex control structures. In a system that allows "abnormal" procedure entry or exit these effects cannot be achieved simply by including appropriate instructions in the procedure body since exit or re-entry may be the result of actions invoked by other procedures, such as a scheduler suspending or resuming a process.

Dlocal expressions can distinguish several different kinds of entry and exit, as described above, and vary their behaviour accordingly. An extreme example would be two Pop-11 processes which run separate Prolog systems which would have to be set up on re-entry to the processes and saved on suspension, all of which could be done within a single Poplog process. In simpler cases, the syntax for dlocal, combined with the fact that Pop-11 procedures have updaters, allows very clear and economical expression. For example a procedure that requires the 15th element of a vector V always to be saved on entry and restored on exit, could simply have the following declaration, which will cause the vector access procedure -subscrv- to be run on entry, and its updater run on exit:

dlocal % subscrv(15,V) %;

The dlocal construct can be used for saving on entry and restoring on exit the value of any expression. Thus even a lexically scoped variable, e.g. declared lexically local to a file or an enclosing procedure, can be treated dynamically by a procedure using it. Most languages that distinguish dynamic and lexical variables treat them as mutually exclusive, whereas Poplog treats the lexical/non-lexical and static/dynamic distinctions as orthogonal.

Additional control facilities are provided by the "continuation stack" and "trail", both required for Prolog. The continuation stack holds "backtrack" information i.e. records describing states of execution that must be restored if a Prolog clause fails. The trail is a stack of references to Prolog variables that have been bound by unification, and thus at some later stage may need to be unbound during backtracking. The continuation stack and trail are located at the base of the call stack, and they can be allocated more space by translating the trail and the call stack in the direction of growth of the call stack.

Although the continuation stack is currently used only by Prolog, it could be used by other languages, and some Pop-11 programmers have invoked the mechanisms directly, as is done in the Pop-11 programs used to implement the Poplog Prolog compiler.

2.3   The High Level Virtual Machine (VM)

Like many other compiler intermediate virtual machines, the Poplog VM is stack based. However, it is higher level than most and thus the production of code for it by a compiler front end is straightforward. Some of the most notable features of this high level virtual machine language are:

- Arithmetic operations do not exist as distinct VM instructions. For example, an addition would be represented as a CALL of the procedure whose name is "+". This is because in Pop-11 new operators and syntactic constructs may be defined (along with their operational semantics) as well as existing ones changed, thus at the source language level no operation can be considered primitive. When producing code for this call at the VM level, the procedure associated with the name "+" will be fetched using the dictionary, and the relevant call planted or inline routine substituted. Note that the dictionary is used only at compile time, not when the procedure call is executed.

- References to addresses do not appear in the VM language. This is due to the fact that most structures are relocatable heap-based objects, and the requirements of incremental compilation. Thus at the VM level, objects are referred to either by name or directly by object references. It is the responsibility of the lower levels of the compiler, using the dictionary, to substitute identifier names with runtime object references.

In total there are 46 VM instructions. To describe them all in detail would require more space than we have, but the following is a brief description of some of the instructions that will hopefully give the flavour of the language.

1.   Stack operations:
    PUSH <word>
        Push the value of the identifier associated with the word <word> onto the user stack.

PUSHS

 Duplicate the item on the top of the user stack.

POP <word>

 Pop the item from the top of stack into the value of the identifier associated with <word>.

2. Procedure calls:

CALL <word>

 Call the procedure that is the value associated with <word>.

UCALL <word>

 Call the updater of the procedure. All procedures may have an updater, e.g. "hd" called normally returns the head of a list, and when called in update mode it updates the head of a list (cf "setf" in Lisp).

3. Conditional and Boolean instructions:

IFSO <lab>

 Jump to the label <lab> if the top item on the stack is not the Poplog item "false". Remove the item from the stack in any case.

AND <lab>

 Jump to the label <lab> if the top item on the stack is the Poplog item "false", otherwise remove the item from the stack and continue. Leaving the item on the stack allows boolean expressions to return values.

4. Directives:

LVARS <word> <idprops>

 Define <word> to be a lexically scoped local variable, with given property <idprops>.

PROCEDURE <props> <nargs>

 Start code generation for a new procedure whose printname is <props> and with <nargs> arguments.

ENDPROCEDURE

 Terminate compilation of a procedure expression and create a procedure record which is then pushed on the user stack.

EXECUTE

 Execute any instructions currently planted at execute level, i.e. commands entered to the top level which are not procedure definitions.

5. Miscellaneous:

LABEL <lab>

 Define the label of the next instruction planted to be <lab>.

FIELDVAL <num> <spec>

 Access field number <num> of object whose structure is defined by <spec>.

2.4   Compilation Using the Two Virtual Machines

In this section we show the relationship between the two virtual machines, the PVM and PIM, by examining what happens when a procedure is compiled using the interactive incremental compiler. There are actually two versions of the PIM known as I-codes ("I" can be taken as standing for "intermediate", "incremental" or "interactive") which are the intermediate form for user (i.e. normal) procedures and M-codes (think of "M" as referring to "machine"). The latter is only used when compiling the sysPOP system sources and will be discussed further in section 3.1. The rest of this section is concerned only with the incremental compiler and I-codes.

The result of procedure compilation is a procedure object containing executable code which will reside in the heap. These procedures can then either be invoked directly by user commands, or by other procedures compiled before or after them. Commands entered at the top level are also formed into heap-based procedures. However these will be executed immediately after being formed, and as they are not referred to by any other object the space used will be reclaimed when necessary by the garbage collector.

The production of a procedure record involves three stages of compilation as was shown in Fig. 6.1. The function of the language-specific compiler front ends is to plant VM code, which they do by calling VM code procedures: there is one procedure for each VM instruction. The method for achieving this varies between the languages, and this is mainly determined by the characteristics of the language. At one extreme there is the ML front end which creates a parse tree for each function definition and then plants VM instructions whilst walking the parse tree. At the other extreme there is Pop-11 whose compiler plants VM instructions as it reads in Pop-11 expressions, and whose only state information is contained in the local variables of the currently active compiler procedures.

The scheme for compiling Pop-11 to VM code could be very restrictive on the syntax permitted, and so a one-deep VM instruction buffer is maintained which allows for reinterpretation of the source code based on context. Thus when compiling the source code for:

a * b(1) -> c;
    ;;; multiply "a" by "b" applied to 1 and
    ;;; assign result to "c"

at the point that the opening parenthesis is found, the VM code emitted is:

```
PUSH "a"
PUSH "b"
```

However, the presence of the opening parenthesis indicates a procedure call, so the second instruction (still in the buffer) is retracted, and the code planted after the closing parenthesis is found is:

```
PUSH "a"
PUSHQ 1
CALL "b"
```

To allow lexically scoped procedures, a stack of procedures in compilation is maintained. When an enclosed procedure definition ends, the state of compilation of the enclosing procedure is restored and compilation of the outer procedure can continue.

As each VM procedure is called, the corresponding VM instruction is appended to the list of instructions being compiled for the current source procedure. The VM code produced by the front ends will obviously be inefficient, and so the VM code must be optimised, but rather than having this as a separate pass which would slow down the compiler, it is performed as the code is planted. For example, when a user stack POP instruction is planted using the POP procedure, the compiler will check for a preceding PUSH, and if found the PUSH will be overwritten and the two instructions replaced with a MOVE instruction.

There is no MOVE instruction at the PVM code level however, but such instructions are available in the PIM. The incremental compiler translates programs to the PIM via instructions referred to as I-codes. There are about 57 I-codes, some of which play an intermediate role in the sense that they are transformed into lower level I-codes during compilation. Some I-codes represent specialisations of the VM instructions which are differentiated according to their actual arguments, e.g. I_CALLPQ for a CALL to a constant procedure (and thus the type of the called object need not be checked). Other I-codes exist for non-checking structure access and arithmetic, procedure prologue and epilogue code, multiway branches, etc.

The type of optimisations performed when the PVM instructions are called are:

• Removing redundant stack usage, e.g. replacing a "push" followed by a "pop" with a "move".

- Replacing certain procedures with inline code, e.g. non-checking integer arithmetic and non-checking object field access.

- Optimising conditional and boolean expressions to use inline comparisons.

- Collapsing branch chains, i.e. if a branch target is another branch instruction, then the target of the first is the target of the second.

From version 14 a number of compile time flags will be made available to users allowing further optimisations to be made at the cost of safety, for instance turning off procedure entry checks, or backward jump checks.

Upon reaching the end of a procedure definition the result is a list of I-codes together with some extra information such as local and global variables used. At this point it may not be possible to produce code for the procedure, e.g. if it contains a jump to a lexically enclosing procedure, in which case code generation must be delayed until the enclosing procedure has been compiled.

When all such references have been satisfied, the I-code list plus declarative information are passed on to the routine which is responsible for planting machine code in a procedure record. This routine makes two or more passes through the I-code list, and for each I-code calls a corresponding function that will emit binary code for that instruction. The first pass allows labels to be translated into an offset from the start of the procedure, and at this point source language variables are translated to their runtime representations. During the second pass the machine instructions are planted in the procedure record. If there is a choice of instruction branch sizes or displacement address operand sizes, then an extra pass will be required if assumptions about the required size are proved incorrect in the first pass.

The dependence of displacement address operand size on the procedure object size is due to the presence of a literal table in each procedure object. This table holds pointers to all the heap-based objects that the procedure references. During garbage collection, the collector can mark from this table, and after compaction the contents are updated to reflect the new state of the heap. Thus all heap accesses performed by the procedure must go indirectly via this table. If heap accesses were done directly then the garbage collector would have to locate and change all embedded pointers in procedures, which would considerably slow down garbage collection.

3. Porting Poplog

Before explaining the work involved in porting Poplog to a new architecture it is necessary to describe how the system code implementing the base Poplog system is compiled using POPC, the compiler for sysPOP system sources. This has to be treated differently from the interactive compiler (a) because, for efficiency, some additional facilities are provided for the dialect of Pop-11 used to implement Poplog and (b) because instead of compiling executable procedure records, the system compiler has to produce files of assembly language instructions that can be used for rebuilding the system. This means that some problems can be left to the assembler and linker provided with the host machine whereas they have to be resolved immediately by the incremental compiler.

Moreover, a requirement for interactive development and testing of programs is that compilation be very fast, whereas re-building a Poplog system can be a slower process, allowing scope for additional levels of optimisation. Other differences are explained below.

3.1 Compilation of System Code

The code production route using I-codes as described in section 2.4 is taken by all user code, including all explicit and autoloadable libraries, most of the Prolog and Lisp compilers and all of the ML compiler. However, when Poplog is re-built or ported to a new machine, the core of the Poplog system is compiled by a different route to form assembler sources which will form mainly static objects at the base of the heap.

There are several differences from compilation of normal user programs:

• As previously explained, an extended language sysPOP is provided for system sources, with its own compiler POPC. POPC is an augmented version of Pop-11, built using the compiler tools available to the user.

• Instead of generating executable procedure records containing machine instructions, POPC outputs assembler files for the target machine (which may be the same as the machine on which it is running or a different machine).

• A different intermediate language, M-code, is used for the low level PIM.

• A different compilation strategy is used to generate M-codes, permitting considerably more optimisation than for user procedures.

- A small subset of the system is written in hand-coded assembler.

- Several of the requirements for system compilation are different from those for normal compilation of user procedures. For example, greater efficiency is required, and in many cases possible, and system procedures and structures cannot be re-located by the garbage collector, so more efficient accessing methods are possible.

The remainder of this section elaborates on these points.

The majority of the system sources are written in sysPOP. This is a dialect of Pop-11 that has been extended to allow the representation of and the operation on: machine integers, addresses, C-like structures, call stack frames, etc. Other additions include only allowing garbage collection and checks for interrupts at specified points, access to operating system calls and calls to assembly language routines.

The compiler for sysPOP, POPC, uses the normal Pop-11 front end (extended to accept the sysPOP language) to produce Poplog VM code. However, rather than this being rewritten down into I-code, the resulting VM-code list is passed almost unadulterated to the optimisation and code generation routines in POPC which generate M-code PIM instructions as follows.

The VM-code list is optimised in a more thorough manner than that for the incremental compiler, and tables of translations between operations (such as machine integer addition) and low level M-code instructions that implement those operations are used to create a list of equivalent M-code instructions.

M-code is similar to the ISP-like "register transfer expressions" that many retargetable compilers use to represent their low-level machine-independent code (Davidson & Fraser [1980]). Some examples are:

M-code	Equivalent register transfer
M_ADD 3 r1 r2	r[2] = r[1] + 3
M_ADD CMP a b EQ lab	if M[a] == M[b] then PC <- lab

There are 47 M-code instructions. A few are more complex than register transfers, e.g. those to create and unwind stack frames on procedure entry and exit respectively and one implementing multi-way branch tables.

The complete M-code list corresponding to a procedure definition is passed to a procedure which steps through the list, and for each M-instruction calls the relevant routine which will emit assembler to implement that instruction. Additional routines also exist to create assembler representations of Poplog objects such as lists and properties.

A small fraction of the system is implemented as subroutines written in hand-coded assembler. These perform operations which are either too primitive or too machine or operating system dependent to be written in sysPOP, or which are so time critical that hand-coding gives a significant overall speed increase.

The first implementation of Poplog was based on the VAX, whose architecture makes it relatively easy for a human programmer to produce good code, and there were quite a large number of hand-coded assembler routines. With a move towards ever simpler architectures, which are harder to program efficiently at the machine level, the sophistication of the sysPOP compiler has been increased, and so more and more of the assembler routines have been replaced by routines coded in sysPOP which are then automatically optimised. However, there remain some which must be written in assembler, for example routines which manipulate the Poplog call stack and therefore cannot be Poplog procedures that would themselves use the call stack.

The hand-coded assembler files are also processed by POPC, and this allows them to use definitions from sysPOP "included" files providing standard definitions and compile time constants, analogous to the use of the "#include" facility in C.

As each source file is compiled, all the names used in the source file, along with their attributes, are recorded in a separate file. When all the source files have been compiled the information about names is processed by a utility called Poplink which is responsible for checking consistency across source files and creating an assembler representation of the Poplog dictionary, and associated word and identifier records. This is necessary because after the new system has been created, access to system identifiers is required at run time, e.g. when user procedures are compiled and refer to system procedures or system global variables. A subset of the names, not required after the system is built, will be marked as not for export, and will not go into this dictionary.

All the assembler files produced by POPC and Poplink are then assembled and linked in the normal manner.

3.2 Work Required to Port Poplog

The architectures to which Poplog has been ported are: DEC VAX, Motorola MC68000 family, GEC Series 63, Intergraph Clipper, Intel 80386 and Sun SPARC. At the time of writing a port to the MIPS processor is nearing completion. Architectures lacking large uniform address spaces (e.g. Intel 80286) are not suitable hosts for Poplog.

The operating systems under which Poplog runs are VMS and several different flavours of UNIX, including Berkeley 4.2, 4.3, SunOS, Dynix, and HP-UX. The four major elements requiring change when Poplog is ported to a new architecture are:

1. System code generator (for POPC, the batch compiler). The routines which produce assembler for M-code instructions must be rewritten. Other changes may be necessary such as producing a routine to emit the binary representation of floating point numbers if a non-standard format is used. In some cases porting involves producing output for a new assembler even though the processor is the same, since different suppliers don't all use the same assembly language format.

2. User code generator (for the run time assembler of the incremental compilers). The routines which produce target binary for I-code instructions must be rewritten for each new architecture. Closures and arrays are held as procedures, and thus routines to plant machine code for them must be rewritten. Arrays could be held as conventional data structures but are held as procedures to make access efficient and also because it is often convenient and elegant to treat arrays as functions.

3. Hand-coded assembler. Roughly 1500 lines of assembler must be rewritten to implement routines which either cannot be implemented in sysPOP or are too time-critical for sysPOP.

4. Operating system interface. Up to 25 sysPOP files may need modifying when porting to a new operating system. The amount of change required depends mainly on the amount of difference between the new operating system and one on which Poplog already runs. (The number of files is larger than might be expected because the system has been deliberately fragmented so that unwanted facilities can be omitted if not required in a "delivery" system.)

The time required to perform the first three tasks can be anywhere between 3 and 8 man months depending on the target architecture and the experience of the implementor. It is not so easy to estimate the time required by the fourth task, but writing for another flavour of Unix would probably take only be a day or so. The initial port to Sequent Symmetry, the first Poplog host with an Intel processor, took just over four months, done by someone who had previously worked on Poplog but had never ported it. After that porting to the Sun386i took about two weeks.

This may seem a long time to perform a port when compared with other systems of comparable complexity that are, say, implemented in C, and can simply be re-compiled, but it must be remembered that porting Poplog includes the production of two back end code generators (tasks 1 and 2). The porting time of a system written in C would not include the time taken to produce the C compiler.

This obviously raises the question why Poplog is not implemented in C. The reasons that this is not so are:

• Various elements of Poplog require direct access to the call stack, which is not possible in C.

• The presence of a garbage collector demands that registers be partitioned into those that can contain pointers to Poplog objects (from which the garbage collector will mark) and those that do not. Such tight control over register usage is not possible in C.

• Poplog requires a global register to point at the top of the user stack. C does not allow global register variables.

• There are benefits in having the system implementation language very similar to Pop-11 such as ease of migrating programs between user and system code, and implementors having one less language to learn.

• The POPC compiler performs optimisations suited to Poplog that could not be expected of a C compiler.

The ease of porting Poplog to an architecture and the quality of the code produced depend on the presence or absence of features in that architecture. Some machine features and their impacts on portability and efficiency are listed below.

• All current Poplog hosts have byte addressing. Whilst a machine without this can support Poplog (indeed, in sysPOP, pointer and machine arithmetic use different operators even though the actions are equivalent for existing machines) the resulting system may be less efficient, e.g. because accessing individual bytes will be more complex.

• Early Poplog ports were all to machines with about 16 registers. Having too small a number of registers (e.g. Intel 80386) can preclude the advantage of register based variables, whereas having a large number of registers (e.g. AMD Am29000) would cause problems in trying to use all the registers effectively.

• A significant proportion of the system's time is spent performing user stack operations and tag manipulation, thus the availability of facilities such as autoincrement/decrement addressing mode (or stack operations using a general purpose register as the stack pointer) and quick forms of arithmetic and logical operations are important.

• Both of Poplog's low level virtual machines have three-operand instructions. Implementations on a two operand machine, which necessarily destroy one source operand, must introduce extra code to save values. An orthogonal instruction set can reduce this problem, and generally makes all code production easier.

• Heap-based procedures must be position independent. Machines without PC-relative branches and data access modes therefore require a procedure base register to be maintained at extra cost. Alternatively garbage collection procedures have to be far more complicated as procedure records will need to be altered when they are re-located.

• Pipelined machines often have delayed branches (the instruction following the control flow change is always executed before the branch target instruction is executed) or delays between loading a register from memory and the point at which the register contents are valid. To produce good code for these machines requires careful scheduling of instructions.

4.   THE COMPILER TOOLKIT

A collection of built-in procedures give users tools for building new incremental compilers, or extending the syntax of the existing languages. These tools are also used for the development of Poplog itself including the implementation of the four Poplog languages.

The lowest level of access, and the one that all compilers will use, are the Poplog VM code planting procedures that were described in section 2.4.

Languages created by extending Pop-11 use two main mechanisms: macros and syntax procedures. Macros are words that when encountered by the itemiser will read in source program text from the input stream and then replace it with modified Pop-11text that is subsequently compiled in the normal way. Syntax procedures are what drive most of the compilation of a Pop-11 program. For example, the "if...endif" construct is compiled to VM code by a syntax procedure whose name is "if". Whenever a word is encountered in the sourcecode stream whose value is a syntax procedure, that procedure is invoked to handle the compilation of the construct. The provision of user-definable syntax procedures, unique to Pop-11, allows a wider range of language extensions than macros, since macros must produce code that is legal according to the rules of the language, whereas syntax procedures need only plant legitimate sequences of VM instructions.

Several languages have been implemented in this fashion including sysPOP, the extended dialect of Pop-11 accepted by POPC, the compiler for the system source code of Poplog. To help in defining new syntax procedures, the user is given access to a collection of procedures which recognise and produce VM code for certain syntactic units of Pop-11 such as expressions, statements, sequences of statements, etc.

Of more general use is the itemiser which has procedures for returning the next item from the compilation stream, checking that an item is present in the stream and replacing an item on the stream by invoking a macro. The itemiser itself is modifiable through the use of character class tables. Thus, for example, one could define "$" to be of class alphabetic, which would then allow "foo$baz" to be recognised as a word. One of the character types, which the Pop-11 compiler assigns to the underscore character, allows alphanumeric characters to be combined with characters of other types, e.g. "class_=".

To implement a new language it is not necessary and sometimes not possible to use the built in mechanisms for analysing the input stream. However, Poplog allows a more conventional parser to be implemented which then plants VM code whilst traversing the parse tree as is done in the case of ML. One of the existing languages could be used to write this, and Pop-11 has proved very suitable for this purpose.

5. THE POPLOG ENVIRONMENT

Poplog provides a host of facilities for teaching and research, and for developing and testing versatile professional software. Examples of such facilities include the following.

• Macro definitions allow conditional compilation, "include" files and textual substitution. As well as aiding software development, command languages can be implemented using this facility.

• A library mechanism allows both explicitly loaded and autoloadable libraries. The latter are automatically loaded when the compiler encounters a name that is not currently defined. This allows systems to be kept small (unnecessary procedures are not loaded) but does not force the user to remember to load a long list of libraries that he requires, as this mechanism is transparent. The list of directories to search for both types of libraries can be modified if required, facilitating user-specific or group-specific extensions to the system.

• Interfaces to operating system facilities, dynamic store management, garbage collection, and a rich variety of data-types including indefinite precision integers, ratios, complex numbers, bit vectors, byte vectors, hashed property arrays, and so on, are all automatically provided, and may be accessed by a new compiler via procedure calls to Pop-11 system procedures.

• There are two garbage collection mechanisms. When sufficient memory and/or swap space are available a "stop and copy" garbage collector can be most efficient. However, if there is not enough space available the non-copying garbage collector is invoked automatically. Users can specify that only the copying garbage collector is to be used, and can control garbage collection by directly invoking it at appropriate times, and by locking the heap to reduce the amount of copying required.

• Poplog contains an integrated editor, VED, which can do much more than just edit files. Programs can be compiled from the edit buffer comprising the whole file, a marked range or the procedure containing the current cursor position. On error the cursor will be positioned at the point of error. Programs can also be executed in an edit buffer allowing the examination of output and the reentering of commands.

• A VED buffer is a datastructure that is readable and writeable by both users and programs, with all changes immediately visible on the screen. This enables the editor to be used as a general purpose, terminal independent, user-interface mechanism. This is not so easy when the editor is a separate process communicating with the AI language system via a pipe. Among other things, the editor has been used to provide an electronic mail interface, an electronic news-reader and a simple character-based graphical program e.g. for displaying parse trees etc.

• On-line help and tutorial files exist for the current system and can be inspected using VED, as can program library files. With the aid of search lists, these can readily be augmented by the individual user or by a teacher for a group of students. Because the editor is integrated with the system, it is easy to include executable examples in the teaching files, including examples that users can modify and experiment with.

• Large programs, once debugged, do not have to be recompiled each time Poplog is restarted, but rather "saved images" can be made. Moreover saved images can be "layered" so that a single saved image can be shared by a group of users, and several different saved images created relative to it, containing code compiled after it is run. This is typically how different Poplog languages are implemented as shareable saved images.

• The efficiency of Poplog as compared with some interpreted AI language systems allows it to be used for a whole range of system development, in addition to AI applications. For example a user developed a fast troff previewer written in Pop-11, enabling formatting to be checked without wasteful printing. For procedures requiring even greater efficiency, e.g. for AI vision research, an interface to C or Fortran is available. A recent change to make Poplog pointers address the first data item after the key will simplify this interface (see Fig. 7.3).

• A "lightweight" process mechanism is provided, which can be used in conjunction with timed interrupts to simulate multiple processes. In conjunction with the section mechanism it allows two Prolog processes to be run in tandem, sharing datastructures.

• The Poplog "Flavours" library implements a powerful object oriented programming system, including demons and multiple inheritance.

• There is a Poplog window manager for certain workstations. This is being replaced by an interface to the X Window System, in order to reduce machine dependence.

• For expert system development, toolkits implemented in Poplog are available.

• In future the sysPOP dialect and POPC compiler for it will be made available to users. This "delivery" mechanism will enable them to build reduced versions of Poplog containing only what is required for their applications. It will also be possible to compile user programs to object modules which can be linked as required in combination with other programs. In some cases it will permit cross-compilation.

All the main features are automatically made available for any new language added to Poplog. Moreover, when Poplog is ported to a new machine, all the development facilities are immediately available, and because Poplog is so portable users are not locked into specialised machines but can take advantage of the increased speed and low cost provided by new general purpose computers, for instance RISC-based workstations.

6. CONCLUSION

Although all the mechanisms described above work on a range of machines and operating systems, there is a price that is paid for the flexibility and generality of Poplog as compared with a single-language system. A compiler and store manager dealing with only one language can often make optimising assumptions that are not valid for a mixed language system. For example in a pure Prolog system programs can be optimised to run faster than Poplog Prolog which has to allow for mixed language interactions.

However, this is counterbalanced by the fact that Poplog Prolog users have the option to identify key portions of their programs that could be re-written in Pop-11, often thereby achieving far greater speed increases than using a stand alone Prolog system. Similarly, programs written in a stand alone Lisp system will often run faster than programs implemented using a more general virtual machine. However, most of the commercially available Common Lisp systems require considerably more memory than Poplog Common Lisp, which initially requires a process size of under 2 Mbytes—including the editor and the Pop-11 system as well as Common Lisp. For those using only Pop-11 and the editor the initial size is well under a megabyte on most machines. One reason for the compactness is that many facilities are provided in autoloadable libraries instead of the main system, which means that users are not encumbered with facilities they are not using.

Pop-11 programs in Poplog seem to run at speeds comparable to programs written using the best Common Lisp compilers. Poplog Common Lisp is not yet (Poplog V13.6) as fast, but further optimisations should remedy this, and in any case the small size can make programs much faster simply by reducing the need for paging.

The fact that the Poplog VM uses a stack for passing arguments and results, allows procedures to take variable numbers of arguments and return variable numbers of results. But this means that it is not generally possible to use optimising techniques that transfer arguments and results via registers. However, the fact that modern processors have increasingly large caches implies that if the top elements of the stack are accessed frequently then that portion of memory will be cache resident. This should considerably reduce the overhead resulting from use of a stack, without any complicated and time-consuming compiler optimisations being required.

The success of Poplog's two-level architecture is attested by the following facts:

• Its use for AI teaching, research and development is continuing to grow, in the UK and elsewhere. For instance, in one UK university it is used in six different departments and in another it is used in four different schools. At the University of New South Wales, in Australia it is the basis of a new MSc degree in cognitive science and is used for teaching undergraduates at Griffiths University.

• It is being used in both commercial and academic organisations.

• Its use for AI purposes ranges over such things as speech and image analysis, natural language processing, theorem proving and expert systems, including real-time expert systems. Commercially available tools have been implemented using it.

• Its use is not restricted to AI — e.g. it is being used to develop a variety of software tools such as text-editors, mail front ends, previewers, compilers for new languages, and the ML system is being used for teaching and research in computer science and software engineering.

• A commercial organisation (PVL) is marketing a program validation system based on Poplog.
• It has proved comparatively easy to add incremental compilers for new languages, e.g. the ML compiler, without having to produce a new development environment for each one including editor, help system, etc.

• It has proved relatively easy to port to new architectures. E.g. the initial SPARC (Sun4) port took under three man-months.

The development and maintenance of the whole system, including all four compilers the editor and the window manager, with implementations on a range of machines and operating systems, has required far less effort than would be required for four separate language systems. The team at Sussex University responsible for Poplog has varied between four and eight full time programmers, with part time help from a few others.

For some users a Lisp machine, with its high performance processor and more sophisticated Lisp development environment might be preferable to Poplog. However, for those with more limited resources and more varied needs, a portable multi language system that runs on generally available hardware can be more attractive. For some University departments and some commercial users, Poplog has played an essential role in making it possible to get on with teaching, research and development in AI. In particular, for experienced programmers in languages like Pascal and Fortran, the Pop-11 sub-language of Poplog has proved a convenient bridge from more conventional languages to AI languages.

7.   ACKNOWLEDGEMENTS

This paper was written (in 1989) by the first two authors, but the main architect of the two-level virtual machine architecture was the third author. Steve Hardy contributed to the initial design of a common virtual machine. Staff and students at Sussex University designed and implemented other aspects of Poplog, and Integral Solutions Ltd ported it to some of the host machines. The Pop-11 language and some of the core features of the Poplog system owe an intellectual debt to the developers of Pop2, Pop-10 and WPOP in Edinburgh, including Robin Popplestone, Rod Burstall, Julian Davies, Ray Dunn and Robert Rae.

The work was supported in part by grants from the UK Science and Engineering Research Council and the Alvey Directorate.

UNIX is a trademark of AT&T.
CLIPPER is a trademark of Intergraph Ltd.
VAX, PDP11 and VMS are trademarks of Digital Equipment Corp.
68000 is a trademark of Motorola Semiconductor Corp.
80386 is a trademark of Intel Corp.
SPARC and SunOS are trademarks of Sun Microsystems.
X Window System is a trademark of the Massachusetts Institute of Technology
POPLOG is a trademark of the University of Sussex

8. REFERENCES

J. W. Davidson and C. W. Fraser: The Design and Application of Retargetable Peephole Optimiser" in *ACM Trans. Prog. Lang. & Sys. vol. 2*, no. 2, April 1980, pp. 191-202.

Aaron Sloman and the Poplog development team: "Poplog: a portable interactive software development environment", Cognitive Science Research Paper No 100, School of Cognitive and Computing Sciences, University of Sussex, 1988.

T. B. Steel, Jr.: UNCOL: the Myth and the Fact" in *Ann. Rev. Auto. Prog.* Goodman, R. (ed.), vol. 2, 1960, pp 325-344.

A. S. Tanenbaum, H. van Staveren, E. G. Keizer and J. W. Stevenson: "Practical Toolkit for making Portable Compilers" in *Communications of the ACM vol. 26*, no 9, September 1983, pp. 654-662.

CHAPTER 7

Applications of Expert Systems Technology to Medicine

Mario Stefanelli
Dipartimento di Informatica e Sistemistica, Università di Pavia, Pavia, Italy

Although much of the technology underlying expert systems available today is derived from basic research on medical advice systems during the 1970s, only two rather small systems, for pulmonary function test evaluation (Aikins et al., 1983) and for interpretation of serum electrophoresis results (Weiss et al., 1983), are in routine clinical use. Medical expert systems will begin to appear, however, as researchers in medical artificial intelligence (AI) continue to make progress in key problems such as general system architectures, management of uncertainty, knowledge acquisition and encoding, man-machine interface and system integration into clinical environments. It is accordingly important for researchers to increase their efforts and for physicians to understand the current state of such research and the theoretic and logistic barriers that remain before useful systems can be made available. This chapter addresses such problems illustrating the most convincing approaches followed today by researchers to provide efficient solutions.

1. ARCHITECTURES OF EXPERT SYSTEMS

An expert system architecture specializes common AI problem-solving techniques to a particular class of tasks. An architecture provides descriptions of a particular kind of problem (e.g., diagnosis, therapy planning, and patient monitoring) at a conceptual level that is above the implementation, thus making clear which aspects of a class of problems are intrinsic to the problem and which are the artifacts of the implementation. An architecture represents a partial design of a knowledge system in which some decisions are made in advance to support specific task characteristics.

For example, most medical diagnostic systems first interpret data bottom-up to find "triggered" disease hypotheses, then set up top-down goals to acquire evidence pro and con the triggered hypotheses. A more sound epistemologic model describes diagnostic reasoning in terms of the three classical types of inferences: abduction, deduction, and induction. Disease hypotheses are abduced from patient data and expected manifestations are deduced and matched against observed ones in order to plan the acquisition of new data, to rule out by induction the unplausible hypotheses and to obtain the best explanation for patient data.

The idea of an architecture level underlies recent work on knowledge systems. Chandrasekaran (1987) has identified a number of "generic tasks" such as hierarchical classification, hypothesis assembly, hypothesis matching, and database inference. Each task is associated with a certain types of knowledge and a family of control strategies. At each stage in the reasoning, the system will engage in one of the generic tasks, depending upon the knowledge available and the state of problem solving.

Clancey has described in detail the heuristic classification method embodied in the HERACLES architecture (Clancey, 1986). In HERACLES inference procedure are represented as abstract meta-rules, expressed in a form of the predicate calculus, organized and controlled as rule sets. A compiler converts the rules into Lisp code and allows domain relations to be encoded as arbitrary data structures for efficiency. The combination of abstract procedures and a relation language for organizing domain knowledge provides a generic framework for constructing knowledge bases for related problems in different domains and also provides a useful starting point for studying the nature of strategies.

Keravnou and Johnson (1986) have argued that the more closely an expert system captures the underlying model of competence, the more satisfactory it will be as a component in a human-computer system. The computer system will be more flexible in its problem solving abilities, will be able to explain and justify its decisions, will be more readily adaptable to a teaching aid, knowledge revisions will be facilitated and a natural dialogue structure will be achieved.

Gruber and Cohen (1987) have developed a strategy for knowledge system design to facilitate knowledge acquisition: they viewed the problem in terms of the incongruity between the representation formalisms provided by an implementation (e.g., production rules, frames, semantic nets) and the formulation of problem-solving knowledge by experts. Therefore, the architecture of a knowledge system can be designed to facilitate knowledge acquisition by reducing representation mismatch. This strategy has been applied in the design of an architecture for a medical expert system called MUM.

Fox et al. (1987) are developing The Oxford System of Medicine (OSM) which would be a computerized information system designed to provide information on a wide range of clinical topics, and able to assist general practitioners in making a variety of medical decisions. Since the first prototype was completed at the end of 1986 the authors of OMS have turned to reconstructing the system and formalizing the decision making and control mechanisms. The main themes of the reconstruction are the specification of system tasks as meta-level knowledge and the clarification of the logic of decision making as distinct from domain specific and quantitative aspects.

All these research projects address the following basic issues:

1. Medical knowledge and reasoning procedures should be represented separately and explicitly;
2. Architectures are not arbitrary combinations of structural components, such as knowledge representation formalisms, inference mechanisms, and control strategies, but artifacts designed by the knowledge engineer for particular tasks;
3. Architectures present representational primitives above the level of their implementation, that is they provide a language describing the behavior of a system in terms natural for the knowledge engineer and expert.

Architectures can also be viewed as epistemological models of medical reasoning and under this perspective they point out the degree of cognitive emulation the designer is pursuing in developing a system. Although there are arguments for and against cognitive emulation as a design strategy attempting to model system performance on human thinking (Slatter, 1987), cognitive emulation seems inherent in knowledge engineering. I am presently more attracted by potential usefulness of cognitive emulation for knowledge engineering than frustrated by its limitations. In the longer term, the combination of developments of Knowledge Engineering and Cognitive Science, the need to tackle larger and more difficult problems and the desire to further humanize the user interface will make a strategy of cognitive emulation increasingly attractive, if not essential.

2. THE DIAGNOSTIC TASK

One of the most fascinating observations from studies of clinical cognition was the discovery that expert clinicians generate a small number of specific hypotheses, very early in the diagnostic process, which are continually evaluated, revised, and elaborated during the process of diagnosis until adequate diagnostic understanding is achieved.

Fig. 7.1 shows an epistemological model we recently developed for describing such a complex cognitive process. It provides a general architecture for diagnostic expert systems and it is currently under implementation into the NEOANEMIA system (Stefanelli et al., 1988).

Such a model will provide here a rational framework for reviewing most of the strategies exploited until now for developing the different components of a diagnostic system and for stressing which are the

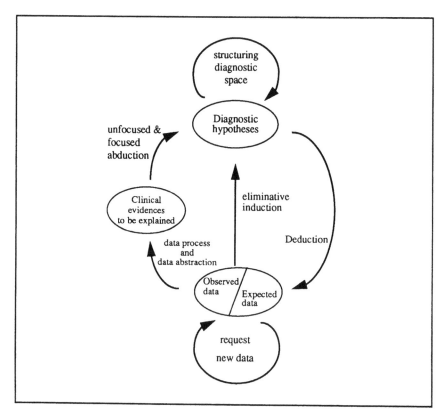

Fig. 7.1.   An epistemological model of diagnostic reasoning.

problems looking for a solution. The following components of a diagnostic system have been identified:

2.1. Data Interpretation

A data interpretation module transforms available patient data into the form used by the rest of the system for its reasoning. This is the process of assessing the validity and meaning of raw input data in the context of the individual case. A very limited effort has been exerted to develop efficient mechanisms to handle validity constraints from pathophysiological knowledge or other fixed bounds, constraints from other parameter values, and the constraints of parameter value changes across time or to handle possibly erroneous data. The long term purpose should be the development of mechanisms able to handle the correction of data.

The second task of data interpretation is assigning meaning to the findings in the context of the patient comparing findings to "normal" could be not sufficient. Normal ranges are determined from the statistics of a population, but the interpretation of a case requires one to establish whether a finding is abnormal in the individual.

Few systems addressed the problems of validity and meaning of data. VM (Fagan, 1980; Fagan et al., 1984) includes a model of the stages a patient follows from Intensive Care Unit admission through the end of the critical monitoring phase in order to interpret the data properly since it depends on knowing which stage the patient is in. For example, when a patient is taken off the ventilator the upper limit of acceptability for expired carbon dioxide measurement is raised.

ABEL (Patil, 1981) interprets electrolyte values in terms of the likelihood they are abnormal as well as in the context of the other electrolytes. If one electrolyte value is changed, it may change the assessment of another electrolyte because the entire picture is now more consistent with a different interpretation. Similarly, the Heart Failure program (Long et al., 1986) uses the distance of a value from the normal range to determine the strength of the evidence for an abnormal value and combines that with the evidences of causes or effects from a physiological model to assess the value.

Within the data interpretation phase can be included what Clancey (1985) called "data abstraction", that is the definition of patient features in abstract terms starting from patient data. Three basic relations for abstracting data have been described: definitional abstraction, qualitative abstraction, generalization. These interpretations are usually made with certainty; belief thresholds and qualifying conditions are chosen so the abstraction is categorical. Data abstraction provides an efficient interface between data acquisition and utilization for reasoning.

2.2. Hypotheses Formulation

Hypothesis formulation through abductive reasoning plays a central role in effective management of the large space of possible diagnoses that a clinician (or a successful AI program) must deal with. The problem of abductive reasoning, as proposed by the philosopher C.S. Peirce (1955), has been a topic of much recent interest in AI (Miller et al., 1982; Reggia, 1983; Charniak and McDermott, 1985; Allemang et al., 1987; Stefanelli et al., 1988). The general task faced by abductive reasoning is to find explanatory useful plausible hypotheses which are worth to be subsequently submitted to a process of critical acceptance where a decision is made as to which explanation is best. Often the term abduction has been used for the whole process of inferring "from data to the best explanation". We reserve the term here for the initial process of hypothesis formulation.

It can be developed using a number of different implementation techniques such as in Present Illness Program (Paulker et al., 1976) by moving frames from long-term memory to short-term memory according to the presence or absence of triggering findings, in INTERNIST-I (Miller et al., 1982) by taking into account the evoking strength associated with each manifestation in a disease profile, in MUM (Gruber and Cohen, 1987) by propagating belief states through the potential-evidence relation within a symbolic inference net or by production rules allowing to infer hypotheses from available data eventually with an associated numeric (Shortliffe, 1976; Miller et al., 1982) or symbolic certainty factor (Quaglini et al., 1986, Torasso and Console, 1988).

Although use of triggers and other associative recall mechanisms provide us with considerable advantage by identifying only those pieces of information potentially relevant to the case, they cannot alone be relied on to reduce the space of possible hypotheses sufficiently for effective diagnostic reasoning. Considerable thought must be given to group and reduce the number of these hypotheses to a small and manageable size for effective diagnostic reasoning. This is one of the goals of the next described diagnostic tasks.

2.3   Structuring of Hypothesis Space

Structuring of diagnostic hypothesis space aims at saving the parsimony of diagnosis through the use of a hierarchy of diagnostic hypotheses and control information defining hypothesis interactions. For example, two diseases might be mutually exclusive (e.g. because they represent distinct sub-types of the same disease), or mutually supportive (e.g. because they are causally associated). INTERNIST-I did not make a clear distinction between hypotheses which are competitors because they are both capable of explaining the same findings in the case (thus not both needed), and those that are competitors because they are mutually exclusive. It was really only concerned with the former type.

Two different research strategies are presently followed to solve efficiently the problem of structuring the diagnostic hypothesis space: developing suitable mechanisms for appropriately treating different types of hypotheses interactions (Josephson et al, 1987) or exploiting a substantially richer and more principled organization of medical knowledge containing a number of hierarchies (Patil, 1981; Pople, 1982; Pople, 1985; Rector, 1987). These hierarchies, however, would be used to organize different components of medical knowledge where each hierarchy deals with a single component for example, one such hierarchy would describe the functional anatomy, another to organize the specific etiologies, and yet others to describe homeostatic mechanisms, temporal characteristics etc. Because of the choice of a single theme for each hierarchy, these hierarchies can be structured in a coherent and systematic fashion such that they provide a smooth and organized progression of concepts as we move from one level in the hierarchy to the next.

2.4. Hypothesis Evaluation

Hypothesis evaluation through deductive and inductive reasoning is essential to complete diagnostic reasoning. Deductive inference is the easiest to explain. Given a hypothesis, such an inference decides which findings have to be expected if this hypotheses is correct. Thus, new laboratory tests or clinical examinations can be requested to test the unobserved expectations deduced from that hypotheses. Inductive inference decides whether hypotheses can be concluded or rejected, or whether they are worth testing further, depending on how closely the observed findings match the hypotheses' expectations. Furthermore, such inference deals with termination of the diagnostic process, that is, it decides whether a "satisfactory" explanation of the patient's problem has been received. This represents a special type of induction, called "eliminative induction", which is quite different from the inductive process able to produce a generalization from a set of observations (Holland et al., 1987).

Two basic implementation techniques for developing diagnostic reasoning can be considered. The simpler one consists in storing expected findings as values of the slots of the frames describing diagnostic hypotheses or in retrieving expected findings by following appropriate paths of a semantic net when this is the formalism exploited for representing medical knowledge. A more promising set of techniques are provided by qualitative reasoning methods (de Kleer and Brown, 1984; Forbus, 1984; Kuipers, 1986; Hunter, 1986; Gotts, 1987; Nicolosi and Leaning, 1987). They provide more expressive power for states of incomplete knowledge than differential or difference equations, by means of a low resolution quantity space representation for values, and classes of monotonic functions for functional relations. Matching and predictions are both facilitated because the landmarks of the quantity space are defined by semantically important points where important changes take place. Current results with relatively small examples have been encouraging, and researchers (Sacks, 1985; Kuipers, 1988; Sacks, 1988) are now taking the steps toward additional mathematical power, hierarchical decomposition, and incremental quantitative constraints, that they believe will make qualitative reasoning essential for developing deep medical reasoning. Much work still remains to be done to extend the possibilities of such techniques, without forgetting that several problems can be tackled more efficiently with traditional numerical modeling techniques when the needed knowledge is available.

The evaluation of the degree to which the observed findings cover the hypothesis' expectations represents another problem looking for general and efficient solution. Most strategies until now proposed assume that patient has one disease. If several disorders are present, the problem is more complex. Additional difficulties arise if the several possible diseases have findings in common or if one disorder influences the presentation of another. The challenge posed by several disorders pushes existing AI programs to their conceptual and computational limits. Nearly all early programs that dealt with several disorders were successful in diagnosing only diseases without overlapping findings. These programs assumed that all hypotheses were competitors and attempted to identify the single most likely diagnosis. Only after the first diagnosis was confirmed did they attempt to make a second diagnosis based on the residual findings. This process needed to be repeated as long as there were findings not accounted for by an already confirmed diagnosis. A major flaw is underlying such a sequential approach: because the program initially has no way of recognizing that more than one disorder exists, findings that are not relevant to the primary disorder can easily confound the diagnostic process.

A partial solution to this problem can be achieved if one assumes that coexisting disorders should, in general, account for a larger set of observed findings than either alone. INTERNIST-I was the first program that exploited this idea. To deal with diseases whose findings overlap or interact represents a more complex problem. In this case the best strategy should be to use pathophysiological knowledge linking diseases and findings through a network of causal relations (Patil, 1981; Pople, 1985; Console et al., 1987). Through this mechanism, which emulates expert human performance, the program can create a composite hypothesis that attempts to explain all of the clinical findings. If several combinations of diseases are consistent with available information, several competing composite hypotheses must be constructed. This process cannot be done in the same fashion as with individual disease hypotheses. Descriptions of individual diseases can be created in advance and made available on demand. Potential composite hypotheses, because they are usually extremely large in number, must instead be fashioned on an individual basis from the findings in a particular case.

A wonderful example of a program that exploits this strategy is ABEL. Its medical knowledge consists of hierarchical representations of anatomical, physiological and etiological knowledge. A disease is then characterized in terms of its anatomical involvement, its etiologic origin, and the functional derangement resulting from it. The disease descriptions are further augmented using causal relationships.

The causal knowledge in the program is organized at several levels of detail. At the shallowest level this knowledge is in terms of diseases and their clinically observable manifestations. At the deepest level this knowledge includes detailed biochemical and pathophysiological mechanisms which provide quantitative relations among normal and abnormal physiological parameters and processes. Additional information is also provided to describe the connection of knowledge at one level to the adjacent levels.

The facts that ABEL is probably the only working program fully implementing pathophysiological reasoning and that further developments have not yet appeared indicate the difficulties in extending such a strategy to other medical problems. These derive from the lack of detailed causal in several medical domains and from the need of developing more general methods for representing multi-level causal knowledge.

The primary advance needed to make the multi-level operators more easily applicable to other medical domains is the ability to define the levels according to the needs in each patient case, rather than relying on fixed levels. To add this capability will require additional operators that will plan the levels for the case by selecting from the knowledge base hierarchy the appropriate specializations of the physiological terms corresponding to the required reasoning level and connecting these together to maintain consistency.

Although ABEL represents an example to be followed by others, I think that the research under development at Ohio University by Josephson et al. (1987) must be taken into serious consideration. They are developing a general problem solving mechanism that is specially suited for performing a particular form of abductive inference, or best explanation finding. A problem solver embodying this mechanism synthesizes composite hypotheses by combining simple hypotheses to satisfy explanatory goals. These hypotheses are formed by instantiating prestored explanatory or diagnostic concepts. In this way the problem solver is able to arrive at complex, integrated conclusions which are not pre-stored.

In conclusion, abductive inference exploits medical knowledge that represents empirically proven relationships between clinical evidences and diagnostic entities, while deductive-inductive inference could make more conveniently use of deep medical causal knowledge. According to the specific medical field, there is a different availability of this second type of knowledge. Moreover, the consideration that the two types of knowledge are often difficult to be fairly sharp separated does not jeopardize the value of their differentiation from an epistemologic point of view: this greatly helps in identifying which generic tasks should constitute the high-level blocks for a medical expert system able to develop diagnostic reasoning.

3.   THE THERAPY PLANNING AND THE PATIENT MANAGEMENT TASKS

Therapy planning and management is one of the least explored areas of medical expert systems. Because of the strong interactions between diagnostic reasoning and patient managing reasoning, integrated systems are needed to adequately face the great majority of real clinical problems.

For example, Digitalis Therapy Advisor (Swartout, 1977), developed at MIT, has represented the first AI attempt to model iterative therapy management behavior, and lead to a program for the diagnosis and management of heart failure because of the need to track the response of the patient's heart failure to the drug. This program managed a patient on digitalis for either atrial arrhythmia or heart failure over the course of a number of sessions. Thus, it encapsulated an iterative approach to finding the patient specific dosage of a drug and made use of a mathematical model of drug kinetics. This last interesting feature will be almost never presented by other programs until the very recent recognition that it is essential in a relatively large number of management problems. Two other therapy management programs were more recently developed at MIT: ABET (Bromley et al., 1983), a program associated with ABEL approaching the problem of symptomatic therapy in electrolyte disorders, and Heart Failure (Long et al., 1986), using a causal physiological model to find and assess therapies as well as diagnose the patient.

While it is clear that therapy planning depends to a large extent on the nature of the diagnostic knowledge representation, there are aspects of the problem that are almost independent. The determination of a therapy plan requires an assessment of the need for the therapy in terms of the risks of the disease state or predicible disease state, the risk and probability of the toxic potential of the therapy, and the probability and potential for benefit from the therapy.

Which techniques are needed to develop an efficient therapy planner and patient manager able to manage properly the above information? First of all, we need flexible and informative implementations of pharmacokinetic models: traditional quantitative techniques need to be combined with qualitative techniques for representing all the available knowledge. These implementations should answer questions about the likelihood of therapeutic and toxic levels, times of high and low drug levels and so forth. Also, we need techniques to implement typical strategies for balancing the therapeutic and toxic risks, and for planning adjustments to the therapy. Three different types of approaches to planning have been used until now: two AI approaches, state space search and skeletal plan refinement, and the decision-theoretic approach.

3.1 Planning Using State Space Search

State space search reduces the problem of planning to the problem of search. These planners typically represent both the initial state and the goal state as predicate calculus assertions (Fikes and Nilsson, 1971). Plan operators are represented in terms of: (i) their preconditions (indicating which assertions must be true of any state to which an operator can apply) and (ii) their effects (indicating which assertions must be added to or deleted from a state each time an operator is applied).

In this framework, the planner searches through the space of all possible states using plan operators to move between states. Several planning strategies are used to aid the search for a path leading from the initial state to the goal state. For example, MOLGEN II (Stefik, 1981), one of the most interesting programs in the domain of molecular genetics experiment design, used constraints and a least-commitment approach to help avoid overcommitting to a particular plan.

State space search planners are effective in using search strategies to find innovative orders of operators to achieve planning goals. Systems of this kind, however, make a number of assumptions which strongly limit their applicability in solving most medical problems. As a matter of fact, the initial state must be known with certainty, the effects of plan operators must be stated explicitly and with certainty, and the goals must be stated such that they can be completely satisfied. Such conditions are seldom satisfied in real problems of patient's therapeutic management.

3.2   Planning Using Skeletal Plans

Skeletal plan refinement is another AI planning technique that uses additional domain knowledge to avoid some of the computational ineffectiveness of state space search. For certain planning problems, plan operators can be grouped into classes that apply at specific points in the problem solving process. This sequence of classes is called skeletal plan, and is used to constrain the search for plan operators. Domain knowledge and strategic knowledge are used to refine the plan operators in the skeletal plan to ones that are appropriate for the problem at hand. Many planning programs, for example VM (Fagan et al., 1980), MOLGEN (Friedland and Kedes, 1985), ONCOCIN (Langlotz and Shortliffe, 1983), and ATTENDING (Miller, 1983), rely on this skeletal plan refinement approach. Because these systems are provided with a skeletal plan in advance, they construct plans that have a consistent overall structure. For example, ONCOCIN is not concerned with when in the plan sequence to give drug therapy, but rather with which of a number of possible variations of drug therapy should be given. Both ONCOCIN and ATTENDING represent their skeletal plans as a hierarchical structure and use additional domain knowledge to refine the plan steps in the skeletal plan.

Skeletal planning is useful when expert guidelines for selecting plan operators can be readily expressed and innovative plan orderings need not be generated (Friedland and Iwasaki, 1985). In addition, skeletal planning systems require that the skeletal plan is refined according to the characteristics of specific planning situations. Consequently, they yield solutions only when the system builder can anticipate the characteristics of most planning problems that are likely to occur. Unfortunately, it is often unfeasible to provide planning solutions for every possible situation in complex medical domains; this would involve specifying knowledge about how to respond to an extraordinary number of specific planning situations.

3.3   Planning Using Decision Theory

The limitations of AI planning techniques described above make clear the importance of representing explicitly both the uncertainty about the patient status and responses to interventions and the tradeoffs among planning goals. A planning technique designed to consider these notions is the decision-theoretic approach for deciding among a group of alternative plans (Raiffa, 1968; Weinstein and Fineberg, 1980). This axiomatic theory combines probabilities and utilities to arrive at a reasoned plan for action. The notion of decision-making by choosing a course of action that maximizes expected utility has been defended by Savage (1972). He proposed the use of subjective probabilities (de Finetti, 1970) to represent uncertainty, and a utility function (von Neumann and Morgenstern, 1953) to represent preferences. The ONYX program (Langlotz et al., 1987) represents an almost unique example of explicitly considering the uncertainties and tradeoffs inherent in all oncology treatment decision.

The basic approach to decision analysis might be characterized as follows:

1. Prepare a structure of a given clinical decision, for example a decision tree, a Markov model, or an influence diagram, which represents the set of candidate decisions and chance outcomes;
2. Assign probabilities to the various possible outcomes;
3. Assign a utility value for each of the possible outcomes;
4. Weight the utilities by the probabilities to score the decisions;
5. Select the decision with the best score as optimal.

In addition, using techniques such as sensitivity analysis it is possible to determine the extent to which assumptions used in a formal model will affect the recommendation that is generated. This in turn helps determine whether further data and analysis are needed before a defensible decision can be reached.

Although the formalism of decision theory offers a comfortable framework for reasoning in narrow domains, its successful applications have been limited to the relatively few groups that are quite experienced with the technology. Ever since the technique was proposed for application to medicine, criticisms have focused on certain obvious problems: the lack of available data, problems in quantifying utilities, the possibility of over-simplification of complex problems, the computational burden of a complex technique, the interpretation of basically quantitative results in a basically qualitative world and the related issue of deciding when a result is significant. These complaints have evoked both epistemologic and technical responses from the clinical decision analysis community.

Probabilities are rarely available for the clinical situation at hand. A critical reading of the clinical literature is often the source of these numbers. Case presentations and analysis in the "Journal of Medical Decision Making" invariably include statements about the literature reviewed and the interpretation of that literature. Nevertheless before the ROUNDSMAN project (Rennels 1986) there was no attempt to build an explicit computational model of how physicians interpret the clinical literature in order to apply the results of biomedical reports to particular cases. ROUNDSMAN represents a first relevant step in exploring how the computer can help to bring a critical analysis of the relevant literature to the physician structured around a particular patient and treatment decision.

Nevertheless the data requirements are no greater than for any form of reasoning: the physician must only admit where knowledge stops and where subjective estimations begins and perform sensitivity analyses to cover his/her tracks. Utilities can be assessed from either physician or patients and formal assessment avoid certain logical limitations of informal reasoning about relative worth.

However in reflecting about the little diffusion of decision theoretic techniques a basic problem emerges. Formal decision theory is far more than a mechanical engine for inference. Its application is complex because it is in a way merely a technique for exploring models that conform to a certain structure. Although far more general its mathematical place in medicine may be no greater than that of specific mathematical models such those describing the kinetics of a drug allowing to locate a patient on an acid-base nomogram or to calculate the valve area in a patient subjected to cardiac catheterization. All such models allow the physician to explore the implications of certain observations to predict the effects of interventions or the effects of future data on his/her reasoning. They facilitate these activities because they make an analogy between the real world and a well understood formal system. Powerful techniques from mathematics are employed to manipulate this formal system; the results may then be transferred by the inverse analogy back to the real-world case hopefully without too much distortion.

The real limitation in the application of decision theory to medicine is that the technique does not encompass the categorical side of medical inference and offers little except formal advice about how useful models can be manipulated. There are four basic activities in which categorical reasoning perhaps even in which expertise and experience are reflected:

1. The structuring of a decision problem;
2. The debugging or validation of a decision model;
3. The interpretation of the results of an analysis;
4. The application of clinical judgment or common sense to ascertain that the conclusions are reasonable.

Thus, much more effort should be devoted to bring the techniques of AI that encompass categorical reasoning on these activities. We need therapy planning systems containing a knowledge base at least on two levels. At level one basic knowledge of the medical domain is required. Level two consists of meta-level knowledge about modeling: what represents coherent and consistent modeling techniques. This is also the level where it is convenient to represent knowledge of the expected behavior of certain kinds of models and knowledge of how certain observed behaviors of a model subjected to sensitivity analysis might indicate specific errors in the underlying problem structure.

Medical planning requires to combine established normative techniques with the symbolic reasoning and the representation techniques developed by AI research. I strongly believe that these combinations are fruitful ones and will likely lead to enhanced decision support for problems in which uncertainty and tradeoffs are dominant problem features.

4. REASONING UNDER UNCERTAINTY

Medicine is characterized by both the knowledge and the data on a new patient being incomplete, the relationships being inexact, and terms being imprecisely defined. These facts raised a long, and sometimes acrimonious debate: is probability theory an appropriate tool for developing systems able to reason under uncertainty? A wide range of diverging opinions can be found in literature (Szolovits and Paulker, 1978; Fox et al., 1980; Spiegelhalter and Knill-Jones, 1984; Cheeseman, 1985; Kanal and Lemmer, 1986; Spiegelhalter, 1986, Henrion, 1987; Clark et al., 1988; Saffiotti, 1988a; Saffiotti, 1988b).

Very briefly, there are four schools of thought. The "logical" model follows the tenets of AI most closely in using only symbolic reasoning and avoiding numerical assessments (Cohen, 1985; Fox, 1986). The "linguistic" model uses fuzzy reasoning to quantify the extent to which the imprecise statements used in common language match formally defined propositions (Zadeh, 1986). A "legal" model uses Shafer-Dempster belief functions to construct arguments for interval-valued belief based on evidence whose reliability is given a numerical assessment (Shafer, 1987). Finally, the "probabilistic" model adheres to the probability calculus, justified both from a theoretical perspective (Lindley, 1987) and from the pragmatic claim that it alone provides flexible and operational means of assessment, criticism and learning (Cheeseman, 1985; Spiegelhalter, 1987).

However, in most implementations of expert systems that acknowledge uncertainty, none of these four paradigms is rigorously adopted. Instead, the somewhat informal numerical schemes used in early systems have been built into shells for general expert system development and hence widely adopted. Hajek (1985) has lately provided a theory for reasonable properties for degrees of certainty attached to production rules. Greater research efforts need to be devoted to the development of systems with the aim of exploring how far handling of uncertainty can be dealt with the alternative approaches mentioned above. Systems like MUM (Cohen, 1987), CADIAG-2 (Adlassnig, 1985), GERTIS (Yen, 1986), and MUNIN (Andreassen et al., 1987) represent valuable experiences in this direction.

Probabilistic coherence is of primary importance in planning therapeutic actions, but it is not always important in diagnostic reasoning. If the diagnostic problem is very complex, exclusively worrying about probabilistic coherence may lead to an underestimation of the importance of the cognitive tasks, and this may result in "shallow" expert systems. In the light of the epistemological model of diagnostic reasoning presented in section 2, do we always require input-output for each reasoning step to be of the probabilistic type, or is it computationally and/or psychologically advantageous to use the logical model providing a categorical output-input? The answer to this question perhaps relies on the notion of utility: in an individual abductive or deductive step there may be little utility in embarking on a coherent probabilistic assessment if conclusions are only temporary. Since a final diagnosis is not usually achieved in a single abductive inference, we can programmatically accept that after each abduction all hypotheses surviving strict exclusionary criteria are "true" with an associated symbolic degree of certainty, because they can have this status modified at a later step, in the light of new evidence on the patient, perhaps containing pathognomonic evidence. Thus, the main effort is not in representing the relationship between a piece of evidence and hypotheses potentially explaining it by means of numerical probabilities, but in assigning to the problem solver the task of deciding what to do when it is uncertain.

Perhaps an outstanding improvement would follow from combining the logical and probabilistic paradigms. For example, if the behavior of the system in a certain subproblem is unsatisfactory, we could translate the relevant portion of the knowledge into a causal network representation and, provided that relevant data are available, use the network representation to optimise the extraction of probabilistic parameters from the data and/or test hypotheses through efficient computational schemes for exact probabilistic manipulations. This requires that the semantics of causal networks would be significantly enriched. For example, concepts of mutual exclusion or complementarity between diagnoses imply symmetric relationships, which are useful in reducing the space of probabilistic parameters since they imply zero conditional probabilities for certain disease combinations (Lauritzen and Spiegelhalter, 1988).

Another fundamental extension of probabilistic networks is towards "temporal reasoning", for example by adopting a probabilistic approach within a state-transition model of the effects of actions (Berzuini et al., 1989). This leads to "Probabilistic Temporal Networks" (PTN) which, beside representing domain knowledge about the rules governing change in the world, provide a computational architecture for reasoning about that knowledge. Using PTN to develop temporal reasoning is facilitated by their isomorphism to Markov chains. They allow both forward and reverse-time propagation of probabilities, and hence provide us with a tool for performing not only prediction, but also explanation and history reconstruction. Moreover, PTN allow us to make variable strength temporal inferences, as well as to reason upon the interpretation of temporal data which may themselves be characterized by variable degrees of certainty. Besides this, they have another benefit, perhaps the most important, namely that of allowing the development of non-monotonic reasoning.

5. ACQUIRING AND ENCODING THE KNOWLEDGE

Exclusively focusing on representing knowledge and using it within a computer would ignore a key additional issue: how the knowledge for a system is acquired and formulated. It is in this area that much of the expert systems mystique has evolved. There is little doubt that the process of mapping the ill-structured knowledge of a medical subspeciality into a form suitable for machine encoding is among the most difficult and time-consuming parts of the expert systems building process. Not only must knowledge engineers be familiar with technical details of the computational tools available, but they must also be willing to make a major commitment to learning enough about the domain of the expert so that discussions of sample problems can be substantive and detailed.

In recent years the knowledge engineering "bottleneck" in expert systems development has encouraged researchers to develop prototype tools that permit experts to "teach" a computer directly about their specialities. The earliest work of this type was a system named TEIRESIAS (Davis, 1979) that allowed infectious disease experts to update and edit the knowledge base for the MYCIN system (Shortliffe, 1976) by critiquing MYCIN's performance on sample cases and entering in English either new rules or modifications to old ones.

The need for such "knowledge editing programs" is now generally recognized. Although such computer-based tools have been developed to assist in either the early formalization of knowledge bases (Boose, 1985; Bennet, 1985) or refinement of relatively mature systems (Davis, 1979; Politakis, 1985), little work has focused on expediting the development of new knowledge bases when the application area is already well conceptualized. This was one of the main lessons learned during the development of ONCOCIN. Building the original version of the system, which contained the knowledge of 23 similar protocols for lymphoma, required nearly two years and some 800 hours of an oncology fellow's time. Adding three more protocols for the adjuvant chemotherapy of breast cancer took several additional months. The length of time required to develop clinically acceptable knowledge bases could be traced to the ambiguity in the protocol documents and the judgmental knowledge with which practicing oncologists tend to supplement the written protocol guidelines. Thus, an interesting knowledge entry program, dubbed OPAL (Musen et al., 1986), for ONCOCIN was developed. OPAL uses computer graphics techniques to display knowledge about cancer treatment plans in a manner intuitive for physicians. Knowledge entered graphically into OPAL is automatically converted by the program to the format used internally by ONCOCIN.

An important design goal in OPAL was to present a model of oncology knowledge intuitive to expert oncologists. In general, knowledge acquisition is facilitated when a knowledge engineer (or computer-based equivalent) can adopt a formal model of the application area that approximates the cognitive performance of the expert in everyday practice. The importance of attempting to match the conceptual model used by the experts can be illustrated by a further example. AI/RHEUM (Kingsland et al., 1986) is a rule-based consultation program for diagnosis in rheumatology, developed at the University of Missouri with the EXPERT knowledge-based system building tool (Weiss and Kulikowski, 1979). The prototype version of AI/RHEUM represented knowledge directly as production rules. The early system was characterized as "completely unworkable because only the experts who had actually worked on the model could understand the logic".

It turns out that rheumatologists often think about diagnosis in terms of major and minor manifestations associated with particular diseases. For example, the diagnosis of rheumatic fever is customarily based on five major criteria and three minor criteria. The presence of either two of the major criteria or one major and two minor criteria indicates a high likelihood of rheumatic fever. The first version of AI/RHEUM was "unworkable" because many physicians could not conceptualize diagnosis in rheumatology in terms of IF/THEN rules. Knowledge acquisition was greatly facilitated when the system was re-written so that diagnostic knowledge could be encoded as "tables" of major and minor criteria.

In a similar manner, OPAL attempted to approximate the mental model used by expert oncologists in thinking about cancer protocols. Although no formal psychological studies were done to elicit conceptual models from physicians, the graphical displays used in OPAL for knowledge entry were designed to mimic the tables and diagrams found in oncology protocol documents. More important the graphical mechanisms provided in OPAL to facilitate entry of existing protocols for ONCOCIN can also be used by oncologists in the design of new clinical trials. OPAL can thus be viewed not only as a knowledge acquisition environment, but also as a tool to assist in the development of new clinical studies.

Recently, Gruber and Cohen (1987) have clearly emphasized how the problem of knowledge acquisition can conveniently be viewed in terms of the incongruity between the representational formalisms provided by an implementation (e.g. frames, production rules, or semantic nets) and the formulation of problem-solving knowledge by experts. Their thesis is that knowledge systems can be designed to facilitate knowledge acquisition by reducing representation mismatch. Principles of "design for acquisition" have been formulated and applied in the design of an architecture for a medical expert system called MUM. It has been shown how the design of MUM made it possible to acquire two kinds of knowledge that are traditionally difficult to acquire from experts: knowledge about evidential combination and knowledge about control.

One of the most critical problems with today's tools of building expert systems is that virtually each new system must be custom-crafted. I have described some of the interesting attempts to speed up this process by developing sophisticated tools for knowledge acquisition and by identifying generic tasks and methods for expert problem solving. However, even as such efforts succeed, they will leave to the system builder the need to gather and formalize the knowledge of the domain of application and to build it into the system. Ultimately, why can't the computer itself play a much more active role in that process of gathering and formalizing knowledge? The problems of "learning" have received a great deal of attention throughout the history of AI research, and there has been a recent revival of interest and active research (Michalski et al., 1986). In addition to the learning approach essentially based on data statistical analysis techniques, there are many efforts to relate learning to structural description of tasks and capabilities of the reasoner, and efforts to describe learning as guided search through a space of possible theories (Mitchell, 1982). Although such attempts are quite promising, their desired effects in facilitating knowledge acquisition in medical domains have still to occur.

6. HUMAN FACTORS ISSUES

There is no set of issues that accounts more fully for the impracticality of most medical expert systems than the failure to deal adequately with the logistical, mechanical, and psychological aspects of system use. Many systems have been designed with an ability to reach good decisions as their primary function. Yet, it has been shown repeatedly that an ability to make a correct diagnosis or to suggest a correct therapy is only one of the performances necessary for system success (Shortliffe, 1982).

6.1. Logistical Issues

Many potential users of expert systems have found their early enthusiasm dampened by programs that are difficult to access, slow to perform, and hard to use. Lengthy interactions, or ones that fail to convey the logic of what is happening on the screen, also discourage use. With hospital information systems increasingly available, it is particularly frustrating if the expert system requires the manual reentry of information known by the user to be available on other computers nearby.

Solutions to such problems mainly require sound institutional strategies for designing a coordinated growth of networking facilities within the hospital or clinic. Similarly, it is a general opinion that expert systems will have their greatest acceptance when such tools are integrated with routine data management functions within a clinic or hospital. As a study of the impact of the introduction of ONCOCIN into the Stanford University oncology clinic showed (Kent et al., 1985), if the physician is using the computer routinely to store and review data, and if that same machine can provide advice that is transparently integrated with its data management function, a major barrier to the use of expert systems may be overcome and the process of data collection may be improved.

Expert systems need to run on powerful enough personal computers that have access to the diverse data and information bases important to decision support. Given the rosy outlook for the expert system market, many vendors are engaged in the continual improvement of existing tools, the migration of the most sophisticated capabilities into smaller and cheaper tools, and the adaptation of tools running on high-cost personal computers such as Lisp machines to more conventional workstations. Noting that the next generation of conventional workstations will have roughly the computing power of today's $50,000 Lisp machines and that several Lisp machine manufacturers have begun to build and market single-chip machines, it seems safe to predict that, by one or the other of these paths, what are today the high-end tools should become available for general use on $5-10,000 workstations.

6.2. Man-Machine Interface

The man-machine interface is known to be an important key of any program's success. Early expert systems exploited only keyboard typing for entering data or answering questions, and most programs require such keyboard-based interaction to this day. I believe that such a requirement can account for much of the resistance to computer use among physicians. Thus, we need to use alternative techniques available that have met with success because they are intuitive to learn and allow the user to avoid typing. Examples include "light pens", "touch screens", and "mouse" pointing devices.

The development of a pointer and icon-based style of interaction begins to be recognized as very suitable styles of interaction for computer-naive users. This style is eminently suited to the wide range of computer-supported tasks that involve a limited number of options: data base management, word processing, spread-sheets, form-based interviewing with menus, pictorial displays of medical information and knowledge. It has revealed the importance of advances in the development of high-resolution bit-mapped displays, which allow at least an order of magnitude more information to be effectively displayed on a screen than was possible in the past using conventional cathode-ray tube screens.

We are also beginning to see early tools for speech understanding; these suggest that physicians may soon be able to talk with computers through a microphone for well-specified tasks, thereby avoiding manual interaction altogether.

In the cognitive aspects of interface technology, significant advances have also occurred. Natural language interfaces to data bases now constitute a well-tried technology, and robust systems are currently commercially available (Harris, 1983). These systems allow a user to retrieve information by posing questions and commands in everyday English. Once installed, these systems are extremely simple to use, and allow even a first-time user in a matter of minutes to get relevant information that previously may have taken several days with a systems programmer as intermediary. Nevertheless, these systems have significant limitations that prevent their immediate extension from natural language interfaces into more general interactive problem-solving systems. Greater advances should occur in the near future in systems' ability to generate well-formed, understandable natural language texts that explain an expert system's conclusions (Swartout, 1985) or explain terms used in the application domain.

Another fundamental issue worth being continuously pursued is represented by the task of modeling users, including attempts to represent users in terms of their previous knowledge, so as to present only as much information as they need and can understand in a comprehensible form. Attempts have also been made to model users in terms of their plans and goals, so as to understand and respond to their needs (Sleeman et al., 1985). These user models correspond to those that people develop and use in their intersection with each other. In fact, many of the pragmatic conventions of human-human interaction rely on such modeling, and normal interactions would go away without it. Thus, I believe that user modeling is an essential addition to human-computer interaction.

6.3. Different Modes for Giving Advice

Most expert systems assume a passive role in giving advice to clinicians. Under this model, the physician must recognize when advice would be useful and then make an explicit effort to access the program that waits for the user to come to it. The physician then describes a case by entering data and requests a diagnostic or therapeutic assessment.

"Consulting model" and "critiquing model" can be considered two different styles of interaction with such systems. In the consulting models, the program serves as an advisor, accepting patient-specific data, asking questions, and generating advice for the user. For example, Present Illness Program (Paulker et al., 1976), MYCIN (Shortliffe, 1976) and INTERNIST-I/QMR (Miller et al., 1986) use the consulting approach.

In the critiquing model, on the other hand, the physician comes to the computer with a preconceived notion of what is happening with a patient or what management plan would be appropriate. The computer programs then acts as a sounding board for the user's own ideas, expressing agreement or suggesting reasoned alternatives. These latter types of systems are difficult to build and are currently available only in prototype form. The first example was ATTENDING (Miller, 1983), a program that critiqued a patient specific plan for anaesthetic selection, induction, and administration after it has been proposed by the anaesthesiologist who will be managing the case. Such critiquing systems meet many physicians' desires to formulate plans on their own but to have those plans double-checked occasionally before acting on them. From ATTENDING a relatively large family of programs have derived: they use the same consultation style and operate in different medical areas with the aim of exploring the effectiveness of the critiquing approach.

Another interesting example was provided by an experimental adaptation of ONCOCIN (Langlotz and Shortliffe. 1983). As originally designed, ONCOCIN was intended to use the consulting model to advise physicians regarding the proper chemotherapy plan for patients being treated on a protocol. The system was later adapted in one experiment so that, if it noted potential problems with a user's own therapy plan, it would initiate a critiquing dialogue regarding alternatives. When the physician's plan was in agreement with the computer's assessment of the situation, however, this version of the program would simply continue its monitoring function and would not interrupt the user who was filling out ONCOCIN's interactive graphical flowsheet.

These examples prove the increasing recognition that expert systems should play a more active role, providing decision support as a byproduct of monitoring or of data-management activities and not waiting for physicians or other health workers specifically to ask for assistance. A great appeal of such systems would be their ability to give assistance without requiring laborious data entry by the physicians themselves because they should be integrated with a comprehensive data base of patient information that has already been gathered from diverse sources within the health care institution. Thus, starting from different considerations, we confirm the validity of a concept previously stressed: the use made of a piece of information is inversely proportional to the effort needed to get it. Even with vast improvements in the quality and range of medical advice that can be provided through advances in expert systems technology, this advice will not be used in daily clinical work until it is rapidly and easily acquirable.

7. VALIDATING EXPERT SYSTEM PERFORMANCE

Validating an expert system's performance is only a part of evaluation, a broader task seeking to assess all expert systems' overall value. In addition to exhibiting acceptable performance levels, expert systems should be usable, efficient, and cost effective, as discussed earlier. Validation is thus the cornerstone of evaluation, since highly efficient implementation of invalid systems are useless.

Separating performance validation from other aspects of evaluation can be difficult. For instance, testing is difficult in the case of systems with poorly designed explanation mechanisms or interfaces.

Validation is often confused with verification. Simply stated, validation refers to building the "right" system, whereas verification refers to building the system "right". Although very important in developing expert systems, here I will not discuss this second issue. Nevertheless, it is worth noting that too little work has been carried out until now on knowledge base verification (Nguyen et al, 1987), and most of this has been done by researchers working on medical problems (Davis, 1976; Suwa et al.,1982).

Typically, engineers have validated expert system performance by running test cases through a system and comparing results against supposed known results or expert opinion. These engineers calculate a percentage for the system's success rate and use subjective judgment to analyze and explain a system's failures where test results contradict known results or expert opinion. An example of this simple approach was the Mycin's early validation (Yu et al., 1979a)

However, it presents several problems. The final percentage obtained hinges on the choice of test cases, and its accuracy hinges on the number of test cases chosen. Moreover, any investigation about inter experts consensus is completely neglected.

Since many expert systems began as research prototypes, validation has often been conducted to qualitatively measure system performance, as with INTERNIST-I, or validation has simply been part of all overall evaluation to assess an expert system's value to a particular domain, as with CASNET (Weiss et al., 1978). However, since medical expert systems must be validated carefully before being used on a regular basis, formal validation methods must be used.

Qualitative validation methods employ subjective comparison of performance. This does not imply that such approaches are informal; we can design highly formal qualitative validation. Where appropriate, qualitative and quantitative methods can be combined. If expert system responses can somehow be quantified, that is expressed in numbers or in ordinal categories, then we can employ suitable statistical techniques to compare expert system performance against either test cases or human experts.

O'Keefe et al. (1987) nicely reviewed the main problems one can encounter in validating expert systems and discussed the basic concepts fundamental to validation. Chandrasekaran (1983), Miller (1986) and Wyatt (1987) examined the evaluation problem in the specific case of medical expert systems.

Until now few systems have been validated using formal methods (Yu et al., 1979b; Hickam et al., 1985; Michel et al., 1986; Alvey et al., 1987a; Alvey et al., 1987b; Quaglini et al., 1988); in all theses cases a Turing test coupled with statistical techniques have been used to blindly validate the systems.

Validation and evaluation require more attention than they presently receive. We need a prescriptive methodology; that is, one explaining how to validate expert systems under certain conditions (such as consultation or monitoring applications) and under certain constraints (such as funding and development time). At present, expert system validation experience is limited. A methodology, or methodologies will evolve only in the light of future collective experience and critical appraisal of that experience.

8. CONCLUSIONS

There are both methodological and practical impediments to further development of medical expert systems. Medicine has historically been a fertile ground for basic research in AI because its problems are highly challenging. Nevertheless, theory continues to be a minor part of efforts in the field. We need to develop further research on the following issues: developing computational models of medical reasoning, developing large-scale and comprehensive systems, incorporating causal knowledge, representing time, dealing more effectively with uncertainty and utilities, and defining formal evaluation methodologies.

As decision analysis and expert systems research appear to be converging, it is important to explore hybrid systems over the near term. Decision models can be expanded to incorporate knowledge bases, causal reasoning, and temporal reasoning. Conversely, expert systems can expand their treatment of uncertainty.

New approaches need to be developed to merge expert systems with the routine aspects of medical information and patient record systems, including, for example, the identification of problems in data surveillance and automated updating of knowledge bases and parameters in decision models. Prototypes of systems that monitor patient data and infer appropriate changes to practice should he developed and validated. Such automated decision-support systems will also need to be adapted to the user, and techniques will need to be developed that determine when the nuances of a problem place it outside the scope of the system.

Although methodological problems are relevant, I believe that practical problems are even more pressing.

There is only a short history of the use of decision-support technology in day-to-day patient and health care management. Vendors are reluctant to add unproven technologies to commercial systems, so development has been limited to single institution research systems. Although there is a great interest in validating other researchers' work, the academic resources available for validating expert systems are small. It may be assumed that the ultimate test of validity of an expert system is its transferability and utility from one clinical setting to another. But such transfer is presently still difficult because of differences in computing and clinical environments.

There are also possible legal impediments to the licensing and use of expert systems in medicine. The Food and Drug Administration has claimed in the USA to be an appropriate regulatory agency for decision technology, but has proposed no formal mechanism of licensure. Although some preliminary essays have been written about the status of expert systems with respect to liability, no precedents have been set that clarify their legal status. When systems migrate from institution to institution, questions of legal ownership and liability also arise. No work has been done on the questions of intellectual property inherent in knowledge bases, nor on the patentability of reasoning strategies.

In spite of the problems mentioned above, advanced medical expert systems are beginning to emerge from research laboratories and are likely to have a profound impact on the way medicine is practiced in the near future. Bringing these systems to a fully developed and clinically validated state will require a major commitment of resources, strong leadership, and interdisciplinary research.

ACKNOWLEDGEMENTS

This work has been supported by grants from M.P.I and National Research Council (grant No. 86.02133). The manuscript reached completion in October 1988 and revised in January 1989.

REFERENCES

Adlassnig K.P.: "Present state of the medical expert system CADIAG-2", Methods of Information in Medicine, 24, 13-20, 1985.

Aikins J.S., Kunz J.C., Shortliffe E. H. and Fallat R.J.: "PUFF: an expert system for interpretation of pulmonary function data", Computers and Biomedical Research, 16, 199-208, 1983.

Allemang D., Tanner M.B., Bylander T and Josephson J.: "Computational complexity of hypothesis assembly", in Proc. of the 10th International Joint Conference on Artificial Intelligence, Morgan Kaufmann, pp. 1112-1117, Milan, Italy, 1987.

Alvey P.L., Myers C.D. and Greaves M.F.: "High performance for expert systems: I. Escaping from the demonstrator class", Medical Informatics, 12, 85-95, 1987a.

Alvey P.L., Preston N.J and Greaves M.F.: "High performance for expert systems: II. A system for leukaemia diagnosis", Medical Informatics, 12, 97-114, 1987b.

Andreassen S., Woldbye M., Falck B. and Andersen S.K.: "MUNIN - a causal probabilistic network for interpretation of electromyographic findings", in Proc. of the 10th International Joint Conference on Artificial Intelligence, Morgan Kaufmann, pp. 366-372, Milan, Italy, 1987.

Bennett J.S.: "ROGET: a knowledge base system for acquiring the conceptual structure of a diagnostic expert system", Journal of Automated Reasoning, 1, 49-74,1985.

Berzuini C., Bellazzi R., Quaglini S. and Stercoli P.: "Probabilistic reasoning about time and action", in Proc. of the 11th International Joint Conference on Artificial Intelligence, Morgan Kaufmann, pp. 528-534, 1989.

Boose J.H.: "A knowledge acquisition program for expert systems based on personal construct psychology", Int. J. of Man-Machine Studies, 23, pp. 495-525, 1985.

Bromley J.H., Patil R.S. and Widman L.E.: "An Approach to therapy formulation for acid-base and electrolyte disorders" submitted to AAA1-83 in the topic area of Expert Systems, 1983.

Chandrasekaran B.: "On evaluating AI systems for medical diagnosis", The AI Magazine, 34-37, Summer 1983.

Chandrasekaran B.: "Towards a functional architecture for intelligence based on generic information processing tasks", in J. McDermott ed., Proc. of the Tenth International Joint Conference on Artificial Intelligence, Morgan Kaufmann, pp. 1183-1192, Milan, Italy, 1987.

Charniak E. and McDermott D.: "Introduction to Artificial Intelligence", Addison Wesley, 1985.

Cheeseman P.: "In defence of probability", In Proc. 9th International Joint Conference on Artificial Intelligence, Morgan Kaufmann, pp. 1002-1009, Los Angeles, USA, 1985.

Clancey W.J.: "Heuristic classification", in Artificial Intelligence, vol. 27, December, 289-350, 1985.

Clancey W.J.: "From GUIDON to NEOMYCIN and HERACLES in twenty short lessons", The AI Magazine, 40-60, August 1986.

Clark D., Baldwin J., Berenji H. , Cohen P., Dubois D., Fox J., Lemmer J., Prade H., Spiegelhalter D., Smets P. and Zadeh J.: "Responses to "An AI view of the treatment of uncertainty" by Alessandro Saffiotti", The Knowledge Engineering Review, 3, 59-86, 1988.

Cohen P.R.: "Heuristic reasoning about uncertainty: an Artificial Intelligence Approach", Pitman, Boston, 1985.

Cohen P.R.: "The control of reasoning under uncertainty: a discussion of some programs", The Knowledge Engineering Review, 2, 5-26,1987.

Console L., Fossa M., Torasso P., Molino G. and Cravetto C.: "Man-machine interaction in CHECK", in J. Fox, M. Fieschi and R. Engelbrecht eds., Proc. of the European Conference on Artificial Intelligence in Medicine, Springer-Verlag, Berlin, pp. 205-212, Marseilles, France, 1987.

Davis R.: "Application of meta-level knowledge to the construction, maintenance, and use of large knowledge bases", Ph.D. Thesis, Stanford University, 1976.

Davis R.: "Interactive transfer of expertise acquisition of new inference rules", Artificial Intelligence, 12, 121-158, 1979.

De Finetti B.: "Theory of Probability", Wiley, New York, 1970.

De Kleer J. and Brown J.S.: "A qualitative physics based on confluences", Artificial Intelligence, 24, 7-83, 1984.

Fagan L.: "VM: Representing time-dependent relations in a clinical setting", Ph.D. Dissertation, Computer Science Dept., Stanford University, Stanford, California, 1980.

Fagan L.M., Kunz. J.C., Feigenbaum E.A. and Osborn J.J.: "Extensions of the rule-based formalism for a monitoring task", in Buchanan B.G. and Shortliffe E.H. (eds), "Rule-Based Expert Systems", Addison Wesley, Menlo Park, California, pp. 397-423,1984.

Fikes R.E. and Nilsson N.J.: "STRIPS: a new approach to the application of theorem proving to problem solving", Artificial Intelligence, 2, 189-208, 1971.

Forbus K.D.: "Qualitative process theory", Artificial Intelligence, 24, 85-168,1984.

Fox J.: "Three arguments extending the framework of probability", in L.N Kanal and J. Lammer (eds.), "Uncertainty in Artificial Intelligence", pp. 447-458, North-Holland, Amsterdam, 1986.

Fox J., Barber D. and Bardhan K.D.: "Alternatives to Bayes? A quantitative comparison with rule based diagnostic inference", Methods of information in Medicine, 19, 210-215,1980.

Fox J., Glowinsky A. and O'Neil M.: "The Oxford System of Medicine: a prototype information system for primary care", in J. Fox, M. Fieschi and R. Engelbrecht (eds.), Proc. of the European Conference Artificial Intelligence in Medicine, Springer-Verlag, Berlin, pp. 213-226, Marseilles, France, 1987.

Friedland P.E. and Iwasaki Y.: "The concept and implementation of skeletal plans", Journal of Automated Reasoning, 1, 161-208, 1985.

Friedland P. E. and Kedes L.H.: "Discovering the secrets of DNA", Communications of the ACM, vol. 28, 11, pp. 1164-1186, 1985.

Gotts N.: "A qualitative spatial representation for cardiac electrophysiology", in J. Fox, M. Fieschi and R. Engelbrecht (eds.), Proc. of the European Conference on Artificial Intelligence in Medicine, Springer-Verlag, Berlin, pp. 88-95, Marseilles, France, 1987.

Gruber T.R. and Cohen P.R.: "Design for acquisition: principles of knowledge-system design to facilitate knowledge acquisition", International Journal of Man-Machine Studies, 26, 143-159, 1987.

Hajek P.: "Combining functions for certainty degrees in consulting systems", International Journal of Man-machine Studies, 22, 59-65,1985.

Harris L.: "The advantages of natural language programming", in M.E Sime and M. Coombs (eds.), "Designing for Human Computer Communication", Academic Press, London, 1983.

Henrion M.: "Uncertainty in Artificial Intelligence: is probability epistemologically and heuristically adequate?", in J. Mumpower (ed.), "Expert Systems and Expert Judgment", Springer, New York, 1987.

Hickam D.H., Shortliffe E.H., Bischoff M.B., Scott A.C. and Jacobs C.D.: "The treatment advice of a computer-based cancer chemotherapy protocol advisor", Annals of Internal Medicine, 103, pp. 928-936, 1985.

Holland J.H., Holyoak K.J., Nisbett R.E. and Thagard P.R.: "Induction: Processes of Inference, Learning, and Discovery", The MIT Press, Cambridge, USA, 1987.

Hunter J.: "Qualitative models in medicine: an artificial intelligence perspective", in Proc. of the 3rd IMEKO Conference on Measurement in Clinical Medicine, Edinburgh, UK, 1986.

Josephson J.R., Chandrasekaran B., Smith J.W. and Tanner M.C.: "A mechanism for forming composite explanatory hypotheses", IEEE Trans. on Systems, Man, and Cybernetics, 3, 445-454, 1987.

Kanal L.N. and Lemmer J. (eds.): "Uncertainty in Artificial Intelligence", North-Holland, Amsterdam, 1986.

Kent D.L., Shortliffe E.H., Carlson R.W., Bischoff M.B. and Jacobs D.D.: "Improvements in data collection through physician use of a computer-based chemotherapy treatment consultant", Journal of Clinical Oncology, 10, 1409-1417, 1985.

Keravnou E. T. and Johnson L.: "Competent Expert Systems", Kogan Page, London, 1986.

Kingsland L.C., Lindberg D.A.B. and Sharp G.C.: "Anatomy of a knowledge based system: Al/RHEUM", MD Computing, 3, 18-26, 1986.

Kuipers B.J.: "Qualitative simulation", Artificial Intelligence, 29, 289-338,1986.

Kuipers B.J.: "Abstraction by time-scale in qualitative simulation for biomedical modeling", in Proc. of the IFAC Symposium on Modelling and Control in Biomedical Systems, Pergamon Press, pp. 158-162, 1988, Venice, Italy.

Langlotz C.P., Fagan L.M., Tu S.M., Sikic B.I. and Shortliffe E.H.: "A therapy planning architecture that combines decision theory and artificial intelligence techniques", Computers and Biomedical Research, 20, 279-303,1987.

Langlotz, C.P. and Shortliffe H.E.: "Adapting a consultation system to critique user plans", International Journal of Man-Machine Studies, 19, 479-496,1983.

Lauritzen S.L. and Spiegelhalter D.J.: "Local computations on graphical structures and their application to expert systems". Journal of Royal Statistical Society, 50, 157-224, 1988.

Lindley D.V.: The probability approach to the treatment of uncertainty in artificial intelligence and expert systems", Statistical Science, 3, 17-24,1987.

Long W.J., Naimi S., Criscitiello M.G. and Kurzrock S.: "Reasoning about therapy from a physiological model", in Proc. of MEDINFO 86, North-Holland, pp. 756-760, Washington, DC, 1986.

Michalski R.S., Carbonell J.G,. and Mitchell T.M.: "Machine Learning", Tioga Publishing Company, 1986.

Michel C., Botti G., Fieschi M., Joubert M., San Marco J.L. and Casanova P.: "Validation of a knowledge base intended For general practitioners to assist treatment of diabetes: a blind study", in Proc. of MEDINFO, North-Holland, pp. 122-127, Washington, DC, 1986.

Miller P.L.: "ATTENDING: Critiquing a physician's management plan." IFEE transactions on Pattern Analysis and Machine Intelligence, 5, pp. 449-461, 1983.

Miller P.L.: "The evaluation of artificial intelligence systems in medicine", Computer Methods and Programs in Biomedicine, 22, pp. 5-11, 1986.

Miller P.L.: "Expert critiquing systems: practice-based medical consultation by computer", Springer Verlag, New York, 1986.

Miller R.A., Pople H.E. and Myers J.D.: "INTERNIST-I: an experimental computer-based consultant for general internal medicine", New England Journal of Medicine, 307,468-476, 1982.

Miller R.A., McNeil M.A., Challinor S.M., Masarie F.D. and Myers J.D.: "The INTERNIST-I/Quick Medical Reference project: status report", The Western Journal of Medicine, 145, 816-822, 1986.

Mitchell T.M.: "Generalization as search", Artificial Intelligence, 18, 203-226,1982.

Musen M.A., Combs D.M., Walton J.D., Shortliffe E.H and Fagan L.M.: "OPAL: toward the computer-aided design of oncology advice system", in Proc. of the 20th Annual Symposium on Computer Application in Medical Care, Computer Society Press, Washington, D.C., pp. 43-51, October 1986.

Nguyen T,A., Perkins W.A., Laffey T.J. ald Pecora D.: "Knowledge base verification", The AI Magazine, 69-75, Summer 1987.

Nicolosi E. and Leaning M.S.: "The use of QSIM for qualitative simulation of physiological systems". in J. Fox, M. Fieschi and R. Engelbrecht (eds.), Proc. of the European Conference on Artificial Intelligence in Medicine, Springer Verlag, Berlin, pp. 73-80, Marseilles, France, 1987.

O'Keefe R.M., Balci O. and Smith E.P.: "Validating expert system performance", IEEE Expert, 81-90, Winter 1987.

Patil R.S.: "Causal representation of patient illness for electrolyte and acid-based diagnosis", Ph.D Thesis, MIT-Laboratory for Computer Science, Techn. Rep. TR-267, Cambridge, Mass, 1981.

Pauker S.G., Gorry G.A., Kassirer J.P. and Schwartz W.B.: "Toward the simulation of clinical cognition: taking a present illness by computer", The American Journal of Medicine, 60, pp. 981-995, 1976.

Peirce C.S.: "Abduction and Induction", Dover, 1955.

Politakis P.G.: "Empirical Analysis for Expert Systems", Boston, Pitman, 1985.

Pople H.E.: "Heuristic methods for imposing structure on ill-structured problems: the structuring of medical diagnosis", in P. Szolovits (ed.), Artificial Intelligence in Medicine, Westview Press, Bouldel, USA, pp. 119-190,1982.

Pople H.E: "Evolution of an expert system: from INTERNIST to CADUCEUS", in I. De Lotto and M. Stefanelli (eds.), Artificial Intelligence in Medicine, North Holland, Amsterdam, 179-208, 1985.

Quaglini S., Stefanelli M., Barosi G. and Berzuini A.: "ANEMIA: an expert consultation system", Computers and Biomedical Research, 19,13-27,1986.

Quaglini S., Stefanelli M., Barosi G. and Berzuini C.: "A performance evaluation of the expert system ANEMIA", Computers and Biomedical Research, 2, pp. 307-323, 1988.

Raiffa H.: "Decision Analysis: Introductory Lectures on Choice Under Uncertainty", Addison Wesley, Reading, Mass., 1968.

Rector A.L.: "Knowledge representation for cooperative medical systems", in J.Fox, M. Fieschi and R. Engelbrecht (eds.), Proc. of the European Conference on Artificial Intelligence in Medicine, Springer Verlag, Berlin, pp. 101-111, Marseilles, France, 1987.

Reggia J.: "Diagnostic expert systems based on a set covering theory", International Journal of Man-Machine Studies, 19, pp. 437-460, 1983.

Rennels G.D.: "A computational model of reasoning from the clinical literature", Ph.D Thesis, Stanford University, 1986.

Sacks E.: "Qualitative mathematical reasoning", MIT-Laboratory for Computer Science, Technical Report TR-329, Cambridge, Mass., 1985.

Sacks E.: "Automatic qualitative analysis of ordinary differential equations using piecewise linear approximations", MIT-Laboratory for Computer Science, Technical Report TR-416, Cambridge, Mass., 1988.

Saffiotti A.: "An AI view of the treatment of uncertainty", The Knowledge Engineering Review, 2, pp. 13-27, 1988a.

Saffiotti A.: "The treatment of uncertainty in AI: is there a better vantage point?", The Knowledge Engineering Review, 3, pp. 87-92,1988b.

Savage L.J.: "The Foundations of Statistics", Dover, New York, 1972.

Shafer G.: "Probability judgment in artificial intelligence and expert systems", Statistical Science, 3, pp. 3-16, 1987.

Shortliffe E.H.: "Computer based Medical Consultation: MYCIN", American Elsevier, New York, 1976.

Shortliffe E. H.: "The computer and clinical decision making: good advice is not enough", IEEE Eng. Med. Biol. Mag., 1, pp. 16-18, 1982.

Slatter P.E.: "Building Expert Systems: Cognitive Emulation", Ellis Horwood, Chichester, England, 1987.

Sleeman D., Appelt D., Konolige K., Riell E., Sridharan N.S. and Swartout W.R.: "User modeling panel", in Proc. of the 9th International Joint Conference on Artificial Intelligence, Morgan Kaufmann~ pp. 1298-1302, Los Angeles, USA, 1985.

Spiegelhalter D.J.: "A statistical view of uncertainty in expert systems", in W. Gale (ed.), "Artificial Intelligence and Statistics", pp. 17-56, Addison-Wesley, Reading, 1986.

Spiegelhalter D.J.: "Coherent evidence propagation in expert systems", Statistician, 36, 201210, 1987.

Spiegelhalter D.J. and Knill-Jones R.P.: "Statistical and knowledge-based approaches to clinical decisions-support systems, with an application in gastroenterology", Journal of Royal Statistical Society, A, 147, 35-77,1984.

Stefanelli M., Lanzola G., Barosi G. and Magnani L.: "Modelling of diagnostic reasoning", in Proc. of the IFAC Symposium on Modelling and Control in Biomedical Systems, Pergamon Press, PP 163-177, Venice, Italy, 1988.

Stefik M.: "Planning with constraints (MOLGEN: Part 1)", Artificial Intelligence, 16, pp. 111-140, 1981.

Suwa M., Scott A. C. and Shortliffe E.H.: "An approach to verifying completeness and consistency in rule-based expert systems", The AI Magazine, 3, pp. 16-21.

Swartout W.R.: "A digitalis therapy advisor with explanations" MIT-LCS Technical Report TR-176, 1977.

Swartout W.R.: "XPLAIN: a system for creating and explaining expert consulting systems", Artificial Intelligence, 21, pp. 285-325, 1985.

Szolovits P. and Paulker S.G.: "Categorical and probabilistic reasoning in medical diagnosis", Artificial Intelligence, 11,115-144,1978.

Torasso P. and Console L.: "Approximate reasoning and prototypical knowledge", to appear in International Journal of Approximate Reasoning, 1988.

Von Neumann J. and Morgenstern O.: "Theory of Games and Economic Behavior", Wiley, New York, 1953.

Weinstein M.C. and Fineberg H.V.: "Clinical Decision Analysis", W.B. Saunders, Philadelphia, Penn., 1980.

Weiss S.M. and Kulikowski C.A: "EXPERT: a system for developing consultation models", Technical Report CBM-TR-97, Rutgers University, 1979.

Weiss S.M., Kulikowski C.A. and Galen R.S.: "Representing expertise in a computer program: the Serum Protein Diagnostic Program", The Journal of Clinical Laboratory Automation, vol. 3, 6, 383-397, 1983.

Weiss S.M, Kulikowski C.A and Safir A.K: "Glaucoma consultation by computer", Computers in Biology and Medicine, 1, 25-40, 1978.

Wyatt J.: "The evaluation of clinical decision support systems: a discussion of the methodology used in the ACORN project", in J. Fox, M. Fieschi and R. Engelbrecht (eds.), Proc. of the European Conference on Artificial Intelligence in Medicine, Springer Verlag, Berlin, pp. 15-24, Marseilles. France, 1987.

Yen J.: "Evidential Reasoning in Expert Systems", PhD thesis, Department of Electrical Engineering and Computer Science, University of California, Berkeley, 1986.

Yu V.L., Buchanan B.G. Shortliffe E.H., Wraith S.M., Davis R., Scott A.C. and Cohen S.N.: "An evaluation of the performance of a computer-based consultant ", Computer programs in Biomedicine, 9, pp. 95-102, 1979a.

Yu V.L., Fagan L. M., Bennet S. W., Clancey W.J., Scott A.C., Hannigan J.F., Buchanan B.G. and Cohen S.N.: "An evaluation of Mycin's advice", Journal of the American Association, 24, pp. 1279-1282, 1979b.

Zadeh L.A.: "Is probability theory sufficient for dealing with uncertainty in AI: a negative view", in L.N. Kanal and J. Lemmer (eds.), "Uncertainty in Artificial Intelligence", pp. 103-116, North-Holland, Amsterdam, 1986.

Author Index

Subject Index